EAI/Springer Innovations in Communication and Computing

Series editor

Imrich Chlamtac, European Alliance for Innovation, Gent, Belgium

Editor's Note

The impact of information technologies is creating a new world yet not fully understood. The extent and speed of economic, life style and social changes already perceived in everyday life is hard to estimate without understanding the technological driving forces behind it. This series presents contributed volumes featuring the latest research and development in the various information engineering technologies that play a key role in this process.

The range of topics, focusing primarily on communications and computing engineering include, but are not limited to, wireless networks; mobile communication; design and learning; gaming; interaction; e-health and pervasive healthcare; energy management; smart grids; internet of things; cognitive radio networks; computation; cloud computing; ubiquitous connectivity, and in mode general smart living, smart cities, Internet of Things and more. The series publishes a combination of expanded papers selected from hosted and sponsored European Alliance for Innovation (EAI) conferences that present cutting edge, global research as well as provide new perspectives on traditional related engineering fields. This content, complemented with open calls for contribution of book titles and individual chapters, together maintain Springer's and EAI's high standards of academic excellence. The audience for the books consists of researchers, industry professionals, advanced level students as well as practitioners in related fields of activity include information and communication specialists, security experts, economists, urban planners, doctors, and in general representatives in all those walks of life affected ad contributing to the information revolution.

About EAI

EAI is a grassroots member organization initiated through cooperation between businesses, public, private and government organizations to address the global challenges of Europe's future competitiveness and link the European Research community with its counterparts around the globe. EAI reaches out to hundreds of thousands of individual subscribers on all continents and collaborates with an institutional member base including Fortune 500 companies, government organizations, and educational institutions, provide a free research and innovation platform.

Through its open free membership model EAI promotes a new research and innovation culture based on collaboration, connectivity and recognition of excellence by community.

More information about this series at http://www.springer.com/series/15427

Nuno Vasco Moreira Lopes

Editor

Smart Governance for Cities: Perspectives and Experiences

 Springer

Editor
Nuno Vasco Moreira Lopes
United Nations University
Tokyo, Japan

ISSN 2522-8595 ISSN 2522-8609 (electronic)
EAI/Springer Innovations in Communication and Computing
ISBN 978-3-030-22072-3 ISBN 978-3-030-22070-9 (eBook)
https://doi.org/10.1007/978-3-030-22070-9

This Springer imprint is published by the registered company Springer Nature Switzerland AG
The registered company address is: Gewerbestrasse 11, 6330 Cham, Switzerland

I dedicate this work to my lovely children, Mariana, Gonçalo and Bárbara. I love you from here to the moon (André Sardet).

Preface

A new governance paradigm is emerging as a result of the rapid urbanization, globalization and technological innovations. The old, sectorial, traditional governance models cannot answer anymore to multidimensional, multidisciplinary and interdependent social, economic and environment sustainability challenges. For addressing these unprecedented challenges, cities all over the world are embracing a new concept to govern their cities; this concept is called smart governance. In this book, we are looking to this smart governance concept in two complementary perspectives, a more theoretical one where the fundamental theory underpinning this concept is identified and deeply described and other more practical one where the application examples of successful smart governance initiatives are given and described in detail. The rationale behind the choosing of those two perspectives was to provide a comprehensive overview of smart governance concept and target the knowledge needs of all stakeholders involved in the construction of a city: governments, industry and civil society. Therefore, we hope this book becomes a useful tool not only for researchers but also for many policy- and decision-makers who are concerned or under the process of transforming their cities in a better place for people living and not only for the present generation but also for future generations. Guarantee the future of the next generations only can be accomplished with sustainable development measures, and smart governance is all about that. In this book, the governments and practitioners can find the most recommendable policies and good practices to endeavour to follow a sustainable development path.

Tokyo, Japan   Nuno Vasco Moreira Lopes

Acknowledgement

I thank the United Nations University (UNU) for giving me all the conditions to start this book and DTx (Digital Transformation CoLab) for letting me continue and finish this work. Thank you very much Joanna for all your support during your stay in the UNU-EGOV.

Contents

Part I Perspectives

1 **Smart Methodologies for Smart Cities:**
   A Comparative Analysis   3
   Nuno Vasco Moreira Lopes and Joanna Rodrigues

2 **Smart City Governance Model for Pakistan**   17
   Nuno Vasco Moreira Lopes and Shahid Farooq

3 **Building a Framework for Smart Cities:**
   Strategy Development   29
   Aroua Taamallah, Maha Khemaja, and Sami Faiz

4 **Social Media as Tool of SMART City Marketing**   55
   Dagmar Petrikova, Matej Jaššo, and Michal Hajduk

5 **Inclusive and Accessible SMART City for All**   73
   Dagmar Petríková and Lucia Petríková

6 **AI, IoT, Big Data, and Technologies in Digital Economy**
   with Blockchain at Sustainable Work Satisfaction to Smart
   Mankind: Access to 6th Dimension of Human Rights   83
   Andrea Romaoli Garcia

Part II Experiences

7 **Modelling of Traffic Load by the DataFromSky System**
   in the Smart City Concept 135
   V. Adamec, D. Herman, B. Schullerova, and M. Urbanek

8 **A Combined Data Analytics and Network Science**
   Approach for Smart Real Estate Investment:
   Towards Affordable Housing 153
   E. Sandeep Kumar and Viswanath Talasila

9 **The City of L'Aquila as a Living Lab:
   The INCIPICT Project and the 5G Trial** 177
   Fabio Franchi, Fabio Graziosi, Andrea Marotta, and Claudia Rinaldi

10 **Mobility in Smart Cities: Will Automated Vehicles
   Take It Over?**... 189
   Ralf-Martin Soe

Index... 217

Part I
Perspectives

Chapter 1
Smart Methodologies for Smart Cities: A Comparative Analysis

Nuno Vasco Moreira Lopes and Joanna Rodrigues

1 Introduction

In 2015, 193 member states of the United Nations signed an agreement in which they assumed the compromise of achieving the 17 Sustainable Development Goals (SDGs) stated in Agenda 2030 [1] until 2030. In the year since, in an event promoted by UN-HABITAT in Paris, 143 countries signed another historical international agreement on climate change [2]. These two international agreements set the base for dealing with the emerging sustainable challenges at international level and national level. To honour the compromises assumed in the scope of these agreements, countries shall now take concrete measures to meet those goals within the defined targets.

In a global panorama where cities are struggling to find a suitable governance model for achieving a short, medium and long sustainable development, the smart city concept has arisen as a solution for addressing those challenges and it seems to be the path that cities are following. However, the concept and what it embraces are far from being consensual and several meanings can be found in literature. The UNU-EGOV defines a smart city as a continuous transformational process, which relies on ICT as an enabler, on the involvement of all stakeholders and on the engagement of citizens towards the quality of life and sustainability. Other similar definitions can be found in the International Telecommunications Union (ITU) and the academy. This lack of a common definition also leads to different visions, goals, objectives and methodologies for implementing a smart city. Without a clear vision and methodology for implementing smart cities it becomes difficult, not to say impossible, to reach the desired goals and sustainability—the point where the

N. V. M. Lopes (✉) · J. Rodrigues
United Nations University, Tokyo, Japan
e-mail: nuno.lopes@dtx-colab.pt; joanna.rodrigues@unu.edu

© Springer Nature Switzerland AG 2020

3

N. V. M. Lopes (ed.), *Smart Governance for Cities: Perspectives and Experiences*,
EAI/Springer Innovations in Communication and Computing,
https://doi.org/10.1007/978-3-030-22070-9_1

problems and needs of present generations are addressed without compromising the needs of future generations.

As far as we found out from the literature reviewed, there is not any scientific publication covering the methodologies for implementing smart cities. The authors made a search for relevant publications on scientific databases with these keywords: "*smart cities*" *AND* "*methodology*" *OR* "*methodologies for smart cities*" *OR* "*smart city methodology*" *OR* "*smart cities methodologies*" *OR* "*methodologies for implementing smart cities*". This search resulted in 29 publications, which in turn have been screened against its titles and abstracts. From this screening process, only five publications were considered relevant to our research (see Table 1.1). The full text of these five publications was read. Paper ID01 presents a methodology for monitoring the city to design buildings and infrastructure better. Paper ID02 describes the best practices for implementing smart and green technologies in cities. Paper ID03 explores key concepts of smart cities and outlines a smart city training program for city leaders and managers. Paper ID04 presents an overview of several evaluation systems and methodologies for measuring urban developments and smart cities. Paper ID05 explores the interaction between citizen, media and city activity, advocating that the interpretation of those interactions can be used to develop city services better, this way improving citizen urban experience.

After reading the full papers, the authors realised that none of them addresses methodologies to implement smart cities as a whole (i.e. with a holistic perspective, considering all its domains); they just propose methodologies to address a small portion, not even a domain, of a smart city. Therefore, this chapter tries to overcome this gap in the scientific literature by making a comparative study of three methodologies which are being used for implementing a smart city as a whole, i.e. considering all its domains. The authors selected Dubai, Istanbul and Montreal from the UNU-EGOV repository of cases because they are from different regions of the world (i.e. the Middle East and North Africa, Europe and Central Asia, and North America, respectively), their methodologies were considered by the research team as good methodologies to take into consideration, and they are very well documented. The remaining part of this chapter is organised as follows: Sect. 1.2 presents the

Table 1.1 List of publications found in scientific databases

ID	Authors	Title	Year	Reference
ID01	Khandokar F., Bucchiarone A., Mourshed M.	SMART: A process-oriented methodology for resilient smart cities	2016	[3]
ID02	Goh B.H.	Smart cities as a solution for reducing urban waste and pollution	2016	[4]
ID03	Kirwan C.G., Yao D., Dong W.	The creative city: An innovative digital leadership program for city decision makers	2016	[5]
ID04	László K.	Indicators to measure urban developments and smart cities	2016	[6]
ID05	Fu Z.	Designing urban experience for Beijing in the context of smart city	2013	[7]

research methodology used to conduct this study; Sect. 1.3 describes the smart city methodologies adopted by Smart Dubai, Big Smart Istanbul and Montreal Smart and Digital City; Sect. 1.4 makes a comparative analysis of these three methodologies to assess their similarities, differences and patterns; and Sect. 1.5 presents the main remarks of this work, its limitations and future research opportunities.

2   Research Methodology

In this section, the research methodology applied in this work is described. To analyse the smart city methodologies adopted by each case we used a case-oriented comparative methodology [8]. Given the multidisciplinary nature of smart cities, which crosses humanities and social and physical sciences, the qualitative method has been used to draw broad, interpretative and critical analysis [9]. The comparison of case studies allows us to emphasise similarities, differences and patterns across cases that share the common goal of being smart. The specific features of each case are described in depth in Sect. 1.3. The rationale for selecting the specific cases has been explained in the previous section. The analytic framework used for cross-case comparison is present in Sect. 1.4. For better understanding the cases and their context, interviews and document analysis were the dominant data collection methods.

The findings from this comparative research will be used for policy formulation and for better understanding of the smart city implementation process. This research work is qualitative; it primarily describes what the phenomena of building smart cities involve.

The research process to conduct this work has five steps, as illustrated in Fig. 1.1, which are described below:

1. *Literature* aims at providing the state of the art on the methodologies to design and implement smart cities.
2. *Methodologies* aims at describing the Dubai, Istanbul and Montreal methodologies for implementing smart city projects in terms of vision, goals and phases.
3. *Analysis* aims at comparing the case studies to identify similarities, differences and patterns in their methodologies and, based on that assessment, propose a methodology for smart cities.
4. *Conclusions* aims at summarising the major findings obtained from this work, identifying its limitations and pointing out future research opportunities.

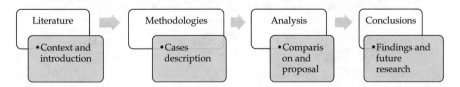

Fig. 1.1 Research methodology

3 Smart City Methodologies

This section describes the methodological approaches followed by Smart Dubai, Big Smart Istanbul and Montreal Smart and Digital City. The methodologies followed by each case study will be described in terms of their smart city vision and the phases they went through to implement it.

3.1 Smart Dubai

Launched in 2014 by Sheikh Mohammed bin Rashid Al Maktoum, Vice President and Prime Minister of the United Arab Emirates and Ruler of Dubai, Smart Dubai is the project of Dubai Municipality (developed by Smart Dubai Office) to transform Dubai into the smartest and happiest city in the world. Using technology as an enabler, 137 initiatives and 1129 smart services have already been rolled out to improve the quality of life of the citizens (Dr. Aisha Bin Bishr, Director General of Smart Dubai Office, personal communication, November 15, 2017).

Smart Dubai's approach began with the definition of the city's vision statement, building on the visionary inspiration of Sheikh Mohammed bin Rashid Al Maktoum:

"Our vision is to make Dubai the happiest city on Earth" [10].

The vision will be achieved by pursuing the mission of embracing technological innovation to make the experience of citizens and visitors efficient, seamless, safe and impactful—these are the four end goals that constitute the strategic pillars of the Smart Dubai project. This clear understanding of where Dubai is heading served as the foundation for developing the whole smart city journey. Figure 1.2 illustrates the sequenced phases of the methodological approach of Smart Dubai.

The next phase of the approach, phase 2, consisted of developing a comprehensive Current State Analysis to assess Dubai's present situation. A series of workshops and questionnaires were carried out with strategic partners to list all existing services and initiatives and to identify challenges. The stakeholders involved in this

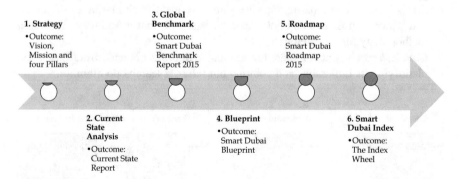

Fig. 1.2 Smart Dubai methodology

part of the process were Dubai Municipality, Dubai Health Authority, Dubai Police, Roads and Transport Authority, Department of Economic Development, Dubai Electricity and Water Authority, Dubai Smart Government, the Executive Council, and Department of Tourism and Commerce Marketing. The resulting Current State Report comprised information on the city's infrastructure, data, systems, applications and electronic services, enabling the identification of opportunities for improvement and facilitating a smooth integration of current and future services to be provided [10].

In the third phase of the smart city methodology, a Global Benchmark procedure took place to identify international mega trends and ICT (Information and Communication Technologies) infrastructure tendencies, originating the Smart Dubai Benchmark Report 2015. First, the Smart Dubai Office evaluated a global smart city research based on benchmark reports from reputable institutions. Then, the top ten cities were selected for further analysis, based on their relevance to Smart Dubai's strategy. After obtaining the list of cities, a series of best practices were identified and ranked according to their relevance to the strategy of Smart Dubai and their innovative character. Finally, the best practices were rated to identify high-potential opportunities for the city to implement [10].

Phase 4 of Smart Dubai's approach consisted of the creation of the Smart Dubai Blueprint, to outline the strategic direction for the participating stakeholders to collaboratively implement the smart city platform. The architecture of the smart city platform was designed, and its four layers (infrastructure connectivity, data orchestration, service enablement and application) and six dimensions (smart economy, smart living, smart mobility, smart governance, smart environment and smart people—smart ICT infrastructure is a transversal dimension underlying all smart services) were defined. Smart Dubai aims to achieve a global leadership position in the smart economy and smart living dimensions, where more opportunities for market development and growth have been identified [10].

Continuing the smart city journey, phase 5 focuses on the roadmap of smart city initiatives and services to be provided in the six dimensions previously mentioned. The Smart Dubai Roadmap 2015 listed all existing and planned services using the data collected for the Current State Report and the Blueprint, then it identified opportunities based on the Global Benchmark Report and assessed all the initiatives and services. Workshops with ten selected strategic partners (government- and private-sector entities) were conducted to build the roadmap of actions cooperatively [10].

The final phase of the process consists of monitoring and measuring the city's achievements. A set of smart city key performance indicators (KPIs) were developed in collaboration with the ITU and the Dubai Statistic Centre to evaluate and measure the level of success of the implementation of the smart initiatives. The Index Wheel of Smart Dubai is a valuable instrument to support decision-making, monitor performance, make necessary adjustments to the process and prioritise specific objectives [10].

3.2 Big Smart Istanbul

Big Smart Istanbul is the project of Istanbul Metropolitan Municipality (IMM) and Istanbul IT and Smart City Technologies Inc. (ISBAK), in partnership with private sector organisations, to conceptualise and implement their smart city vision. Launched in May 2016, the project will start the implementation phase of the Roadmap in March 2018 within a 5-year timeline and several initiatives have already been started (Ersoy Pehlivan, Smart City Coordinator of ISBAK, personal communication, November 15, 2017).

Figure 1.3 illustrates the sequenced phases of the methodological approach of Big Smart Istanbul.

The smart city journey has been developed in five phases. The first phase was dedicated to Literature Review, which comprised a global research review and a Best 10 Smart Cities Analysis. In the global research review stage, a series of smart city domain reports were developed on smart governance, smart energy, smart building, smart mobility, smart infrastructure, smart technology, smart healthcare and smart citizen, and a capacity-building training session on international best practices (with more than 200 participants) was delivered. The best smart cities analysis stage involved three shortlisting procedures to select the ten best smart cities for final analysis, based on criteria such as presence in global indexes, top cities by regions and current smart projects, and a set of diverse indicators (macro indicators: population density and gross domestic product based on purchasing power parity; direct indicators: innovation, public engagement, policy support, vision, breadth and investment; indirect indicators: tertiary education, ease of business, ICT index, industry mix and mobility mix) [11].

Phase 2 was developed in five stages, focusing on the Current State Analysis of Istanbul. In the first stage, a stakeholder management plan was produced, concentrating on 8 domains, 27 focus areas, 61 sub-focus areas and 224 stakeholders (nongovernmental organisations [NGOs], public and private sectors). The next stage was

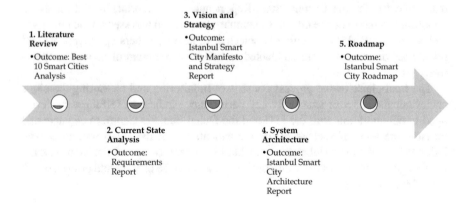

1. Literature Review
•Outcome: Best 10 Smart Cities Analysis

2. Current State Analysis
•Outcome: Requirements Report

3. Vision and Strategy
•Outcome: Istanbul Smart City Manifesto and Strategy Report

4. System Architecture
•Outcome: Istanbul Smart City Architecture Report

5. Roadmap
•Outcome: Istanbul Smart City Roadmap

Fig. 1.3 Big Smart Istanbul methodology

the stakeholder analysis, including online surveys, face-to-face interviews and eight serial workshops for findings and information sharing. Stage 3 was dedicated to the current state report, originating the Istanbul Smart City Index Study and a SWOT (Strengths, Weaknesses, Opportunities and Threats) analysis. In the final stage, gap analysis, a Requirements Report was drawn up, resulting in the comparative analysis between the best ten smart cities and the Istanbul Smart City Index [11].

In the third phase of the smart city methodology, the Vision and Strategy for Big Smart Istanbul were defined. A stakeholder survey was conducted with 203 executives, a social media analysis and persona study were carried out with 450 citizens, a vision search conference was held with 54 participants from various sectors (municipalities, ministries, NGOs and private companies), and a vision assessment conference ("Mars Group Workshop") with 11 participants and special experts previously selected at the previous vision conference was also organised. The next stage built on the data previously collected in other phases was to define focus area strategies—a list of necessities, initiatives and projects for the city. The outcome of phase 3 was the Istanbul Smart City Manifesto and Strategy Report:

> "We believe that everything we do is to protect and improve the city's identity and spirit, prepare for the future, and oversee the rights of past and future generations by recognising the city's own potential and by reproducing with inspiration from the city. Together, we will achieve that through producing with stakeholders and citizens, using technology with innovative methods, and focusing on productivity. We will do so with people who value lore, basic human values, self-confidence, proactiveness and enjoying life. We will provide smart city services and opportunities for all the city stakeholders, knowing today's needs and foreseeing tomorrow's. We will be the top contributor to the quality of life as Smart City Istanbul by 2030" (Ersoy Pehlivan, Smart City Coordinator of ISBAK, personal communication, November 15, 2017).

Phase 4 determined the Architecture of the Smart City System, comprising key stakeholder workshops, conceptual design of the system, high-level technical architecture (covering several layers: technology layer, information layer, application layer, business layer and strategy layer), physical architecture and logical architecture, originating the Istanbul Smart City Architecture Report [11].

Finally, the fifth phase is dedicated to the Roadmap definition and implementation, involving the creation of a governance and public relations (PR) plan, capacity-building training on project finance and feasibility, development of smart city application projects and definition of the project schedule in the short-medium-long term. The Istanbul Smart City Roadmap is the main outcome of this phase [11]. All the initiatives of Big Smart Istanbul are organised under 11 pillars, which in turn are divided into 8 functional areas and 3 enablers. The eight functional areas are smart economy, smart people, smart living, smart governance, smart security and safety, smart mobility, smart energy and smart environment. The three enablers, which are foundational for these eight functional areas, are smart financing, smart organisation and human resources, and ICT (Ersoy Pehlivan, Smart City Coordinator of ISBAK, personal communication, November 15, 2017).

3.3 Montreal Smart and Digital City

In 2014, the City of Montreal launched its smart city initiative, Montreal Smart and Digital City, a collaborative project between institutional and private sector entities, city workers and citizens. The Strategy and Action Plan was designed for the period from 2014 to 2017, investing in innovative growth-generating projects in the areas of urban mobility, direct services to citizens, way of life, democratic life, sustainable development and economic development. The first phase of the smart city methodology in Montreal was defining its Vision:

> "By 2017, Montreal will be the world's leading smart city. As part of this process, the city and community intend to invest in innovative, growth-generating projects. Montreal seeks to devise and develop an outstanding quality of life and a prosperous economy with and for citizens through collaborative innovation, state-of-the-art technologies and a bold approach, backed by Montreal's trademark creativity" [12].

Figure 1.4 illustrates the sequenced phases of the methodological approach of Montreal Smart and Digital City.

This process began with the establishment of the Bureau de la Ville Intelligente et Numérique (BVIN—Smart and Digital City Office), which defined the four focus areas: collection (collecting data for transparent management and open government), communication (access to information and distribution systems), coordination (providing digital public services) and collaboration (supporting industry to stimulate innovation) [12].

Phase 2 was the listening stage of the methodology, focused on identifying needs, problems and priorities through an extensive consulting process involving citizens, city workers and the community while also analysing international models. The BVIN conducted 4 surveys (7601 people responded via Web and telephone) for data analysis, a study of citizen's 311 requests (analysing 1,033,345 calls and 40,000 emails to the 311 service in 2013), a review of best practices in international cities

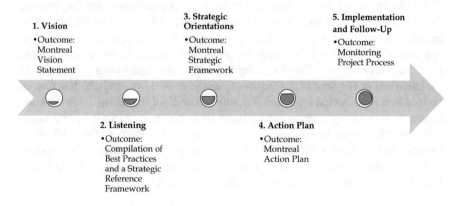

Fig. 1.4 Montreal Smart and Digital City methodology

(originating 7 international case studies focused on the areas of urban mobility, direct services to citizens, way of life, democratic life, sustainable development and economic development), 5 town hall meetings with 203 participants, 2 local activities to co-design public policies with 190 expert participants and an analysis of the ongoing projects and innovation assets in Montreal. The outcomes of this phase included a Compilation of Best Practices and a Strategic Reference Framework [12].

The next phase focused on Strategic Orientations. BVIN's activities and studies revealed five focus areas (economic development, urban mobility, direct services to citizens, way of life and democratic life) and four structural components (telecommunications, open data, technological architecture and community) supporting them, each one with individual goals and relevant international examples [12].

The fourth phase was dedicated to the conception of the Action Plan. The projects are led by the municipality (BVIN), with the help of project promoters, project facilitators and project support. 232 ideas came out of the previous phases, originating prospective projects formulated with stakeholders that started in the short term. Then, major prioritised projects were defined to be developed in the long term, with charters, KPIs and goals. Projects were selected based on their impact on structural components, contributions to strategic orientations, impact on the citizens, cost-effort-return on investment and short- or long-term implementation period. The result of this process was the final Action Plan [12].

Finally, phase 5 is the Implementation and Follow-Up step of the whole process. Each project's progress is carefully reviewed on an ongoing basis and posted to Montreal Smart and Digital City website. Each project goes through four stages: discovery, alpha, beta and fully operational. KPIs are specific for each project and serve for impact measuring and monitoring [12].

4   Comparative Analysis of Methodologies

Each of the smart city methodologies described in the previous section follows specific phases to define and achieve their vision of a smart city.

The phases that comprise the methodological approaches of the three cities are mapped in Table 1.2 to facilitate comparison in terms of similarities, differences and patterns.

After analysing the methodologies used in the process of developing smart city initiatives and comparing their points of contact and disparities, we can draw a few findings:

- The main building blocks of the smart city methodologies are basically the same; only the organisation and the names of each phase are different. The blocks that can be extracted from this comparison are: Vision and Strategy; Current State Analysis and Global Benchmark; Technological Architecture; Action Plan/ Roadmap; Implementation and Monitoring.

Table 1.2 Comparative analysis of smart city methodologies

PHASES	SMART DUBAI	BIG SMART ISTANBUL	MONTREAL SMART CITY
Phase 1	Strategy: definition of the city's Vision, Mission and Pillars. Outcome: Vision Statement, Mission Statement and four Pillars (strategic end-goals).	Literature Review: global research review and analysis of best smart cities practices. Outcome: Best 10 Smart Cities Analysis.	Vision: establishment of the Bureau de la Ville Intelligente et Numérique (BVIN) and definition of the focus areas – collection, communication, coordination, and collaboration. Outcome: Montreal Vision Statement.
Phase 2	Current State Analysis: workshops and questionnaires developed with strategic partners, confirm existing services and initiatives, and identify challenges. Outcome: Current State Report.	Current State Analysis: stakeholder analysis, Istanbul Smart City Index Study and gap analysis. Outcome: Requirements Report.	Listening: identifying needs, problems, and priorities, surveys for data analysis, a study of citizen's 311 requests, a review of best practices in seven international cities, town hall meetings, local activities to co-design public policies and an analysis of ongoing projects and innovation assets. Outcomes: Compilation of Best Practices and a Strategic Reference Framework.
Phase 3	Global Benchmark: evaluate global smart city research, select top ten cities, extract and rank best practices, and identify opportunities for Dubai's smart city initiative. Outcome: Smart Dubai Benchmark Report 2015.	Vision and Strategy: persona and social media analysis, organisation of vision conferences and definitionof focus area strategies. Outcome: Istanbul Smart City Manifesto and Strategy Report.	Strategic Orientations: define focus areas (economic development, urban mobility, direct services to citizens, way of life, and democratic life) and structural components (telecommunications, open data, technological architecture, and community) supporting them, each one with individual goals. Outcome: Montreal Strategic Framework.
Phase 4	Blueprint: design the architecture of the smart city platform, define layers and dimensions, and select participating stakeholders to implement initiatives and provide smart services.	System Architecture: key stakeholder workshops, conceptual design, high level technical architecture, physical architecture and logical architecture.	Action Plan: filter the ideas that came out of the previous phases, define prospective projects for the short-term and define prioritised projects for the longterm.

Table 1.2 (continued)

		Outcome: Smart Dubai Blueprint.	Outcome: Istanbul Smart City Architecture Report.	Outcome: Montreal Action Plan.
Phase 5		Roadmap: list existing and planned services, identify opportunities, evaluate initiatives and services, and organise workshops with strategic partners to collaboratively build the roadmap of action.	Roadmap: creation of governance and PR plan, capacity building training on project finance and feasibility, development of smart city application projects and definition of the project schedule.	Implementation and Follow-Up: review of each project's stage on an ongoing basis and measurement of specific KPIs for each project.
		Outcome: Smart Dubai Roadmap2015.	Outcome: Istanbul Smart City Roadmap.	Outcome: Monitoring Project Process.
Phase 6		Smart Dubai Index: development of a set of smart city KPIs to measure the city's achievements. Outcome: The Index Wheel.	N/A	N/A

Caption

Vision and Strategy	(white)
Current State Analysis and Global Benchmark	(grey 5%)
Technological Architecture	(grey 15%)
Action Plan/Roadmap	(grey 25%)
Implementation and Monitoring	(grey 35%)

- Despite the fact that Dubai, Istanbul and Montreal are located in three different regions (as defined by the World Bank: the Middle East and North Africa, Europe and Central Asia, and North America, respectively), they have similar methodological approaches.
- Vision is perceived as something that should be defined in a primary stage in Smart Dubai and Montreal Smart and Digital City, whereas Big Smart Istanbul only defines the vision in the third phase after reuniting with multiple stakeholders and strategic partners.
- Each city formulates its own proposal for the Technological Architecture of the smart city.
- Regarding the final phase, cities have distinct approaches to evaluate project success: Smart Dubai provides a citywide index for measuring and monitoring

Table 1.3 Proposed methodology for building smart cities

Phase number	Phase name	Step number	Step name
1	Context analysis	1	Benchmarking best practices
		2	Current state analysis
2	Capability assessment	3	Smart city stakeholders
		4	Institutional capacity
3	Strategy	5	Smart city vision
		6	Objectives and governance model
4	Action plan	7	Align initiatives with smart city domains and SDGs
		8	Piloting projects and programmes
5	Evaluation	9	Project monitoring
		10	Impact reporting

impact in all smart city domains, Big Smart Istanbul does not mention this topic in its smart city methodology, and Montreal Smart and Digital City conducts individual project monitoring processes instead of using a holistic approach.

Based on these findings and on the UNU-EGOV-proposed Framework for Smart Sustainable Cities, described in the report Smart Sustainable Cities—A Reconnaissance Study [13], we recommend a five-phase methodology with ten steps to build smart cities, as shown in Table 1.3.

The methodology has five phases: first phase—context analysis, with the steps: (1) benchmarking the best practices of smart city initiatives around the world and (2) current state analysis, taking into account all cultural/social, regional, economic, legislative, infrastructural and technological aspects; second phase—capability assessment, with the steps: (3) smart city stakeholders, their roles and responsibilities and (4) institutional capacity for building smart cities; third phase—strategy, with the steps: 5) smart city vision definition and 6) objectives for the city and governance model to manage the smart city; fourth phase—action plan, with the steps: (7) aligning the initiatives with the SDGs and smart city domains and (8) piloting projects and programmes before scaling up the selected solutions; and finally, fifth phase—evaluation, with the steps: (9) monitoring the progress of the projects and (10) measuring, evaluating and reporting on the impact of the initiatives.

5 Conclusions

Having a well-defined methodology for implementing smart cities is crucial to reach the desired international, national and local sustainable development goals and, as far as we know, this topic is not yet covered in the research literature. Therefore, this chapter tries to give an overview on Dubai, Istanbul and Montreal methodologies, extracting their similarities, differences and patterns. We found out that capability assessment of the involved stakeholders and institutions was not a clear concern in any of the methodologies, which focused mainly on analysing the current state of technology, infrastructure and services without considering the needed capabilities for building a smart city. Based on the findings, and on the experience gathered by the design of the UNU-EGOV Framework for Smart Sustainable Cities, **a five-phase methodology**, *(1) context analysis, (2) capability assessment, (3) strategy, (4) action plan and (5) evaluation*, **with ten steps**, *(1) benchmarking best practices, (2) current state analysis, (3) smart city stakeholders, (4) institutional capacity, (5) smart city vision, (6) governance model, (7) aligning initiatives with SDGs and domains, (8) piloting projects and programmes, (9) project monitoring and (10) impact reporting*, for building smart sustainable cities was proposed in this research work. This study only took into consideration the publications indexed in the Scopus database and the methodologies considered were the ones found by the research team, which does not exclude the possibility that other well-defined methodologies could exist. The proposed and recommended methodology still needs to be tested for validation. In the future, the research team wants to test and validate the methodology with pilot cities and identify the worldwide best practices for each step of the methodology.

Acknowledgements This chapter is a result of the project "SmartEGOV: Harnessing EGOV for Smart Governance (Foundations, methods, Tools)/NORTE-01-0145-FEDER-000037", supported by Norte Portugal Regional Operational Programme (NORTE 2020), under the PORTUGAL 2020 Partnership Agreement, through the European Regional Development Fund (EFDR).

References

1. T.G. Assembly, *Transforming our world: the 2030 Agenda for Sustainable Development— draft* (2015)
2. United Nations, *The Paris Agreement* (2016)
3. M. Mourshed, A. Bucchiarone, F. Khandokar, SMART: A process-oriented methodology for resilient smart cities, in *2016 IEEE International Smart Cities Conference (ISC2)* (2016), pp. 1–6
4. G.B. Hua, *Smart cities as a solution for reducing urban waste and pollution.* (IGI Global, 2016)
5. C.G. Kirwan, D. Yao, W. Dong, *The creative city: an innovative digital leadership program for city decision makers* (Springer, Cham, 2016), pp. 540–550
6. K. László, *Indicators to measure urban developments and smart cities*, vol. 16 (2016)
7. Z. Fu, *Designing Urban Experience for Beijing in the Context of Smart City* (Springer, Berlin, Heidelberg, 2013), pp. 279–288
8. Delwyn Goodrick, *Comparative Case Studies*, vol. 9 (2014)
9. P. Baxter, S. Jack, *The Qualitative Report Qualitative Case Study Methodology: Study Design and Implementation for Novice Researchers*, vol. 13, no. 2, pp. 12–11
10. Smart Dubai Office, Smart Dubai. [Online]. Available: http://www.smartdubai.ae/. Accessed 22 Jan 2018
11. F. Gündoğan, Big smart Istanbul roadmap (2017) [Online]. Available https://pt.slideshare.net/ sitecmy/istanbuls-big-smart-roadmap. Accessed 21 Jan 2018
12. Montreal's Smart and Digital City Bureau, Montreal smart and digital city 2014–2017 strategy (2014) [Online]. Available http://villeintelligente.montreal.ca/sites/villeintelligente.montreal. ca/files/montreal-strategy-smart-and-digital-city-an.pdf. Accessed 23 Jan 2018
13. E. Estevez, N. Lopes, T. Janowski, Smart cities for sustainable development—reconnaissance study (2015)

Chapter 2
Smart City Governance Model for Pakistan

Nuno Vasco Moreira Lopes and Shahid Farooq

1 Introduction

Pakistan, like many other Asian countries, is becoming rapidly urbanized. The urban share of population has increased from 37.9% in 2013 to 40.54% in 2017 [1]. By 2030, an expected 50% of Pakistanis will live in cities, up from the current 40%. Pakistani cities contribute 55% of the country's total GDP [2]. Pakistani cities are a major source of employment opportunities for small- and medium-sized enterprises (SMEs) which provide the vast majority of Pakistan's nonagricultural jobs and high-growth industry jobs such as information technology [3]. Cities in Pakistan are also the hub for Pakistan's educational institutions which impart education, trainings, skills, and research and development opportunities in marketable disciplines [4].

However, Pakistani cities are suffering from many challenges and "without better urban planning to accommodate rapid growth, cities have the potential to become hotbeds of discontent and unrest rather than engines of growth and innovation" [5]. In this context, policy planners at federal and provincial level are aware of the situation. Pakistan Vision 2025 highlights the need of "transforming urban areas into creative, eco-friendly sustainable cities through improved city governance, effective urban planning, efficient local mobility infrastructure and better security." The document envisages the concept of Smart Cities in Pakistan—the cities that are "are digitally connected, equipped with wireless network sensors and there is e-connectivity in all parts where the free flow of information is possible, thereby laying the foundations for the cities of Pakistan to be smart and creative" [6].

On the other hand, these policy documents only hint upon Smart City transformation without going into the dynamics of Smart Cities in Pakistan context, particularly in governance context. However, researchers believe that governance is the key execution

N. V. M. Lopes (✉) · S. Farooq
United Nations University, Tokyo, Japan
e-mail: nuno.lopes@dtx-colab.pt; farrop@unu.edu

© Springer Nature Switzerland AG 2020   17
N. V. M. Lopes (ed.), *Smart Governance for Cities: Perspectives and Experiences*,
EAI/Springer Innovations in Communication and Computing,
https://doi.org/10.1007/978-3-030-22070-9_2

challenge for smart cites [7]. The issues like limited transparency, fragmented account-ability, unequal city divisions, and leakage of resources are some of integral character-istics of regular governance. A move from this type of governance is essential for an effective and efficient administration of the smart cities [8].

Smart governance is an important characteristic of a smart city that is based on citizen participation [9]. Smart governance relies on the implementation of smart governance infrastructure that facilitates service integration, collaboration, commu-nication, and data exchange [10].

2   Smart Governance: Theoretical Overview

Governance is an often-used concept with multiple connotations [11]; however at its root, governance refers to the way in which power and authority are exercised "to manage the collective affairs of a community (or a country, society, or nation)" [11]. Extending the same concept in a smart perspective, many definitions of smart gov-ernance also emerged. Albert Meijer defines smart governance as "using new tech-nologies to improve urban governance through better use of information and better communications" [12]. According to Helmut Wilke, smart governance "is an abbre-viation for the ensemble of principles, factors, and capacities that constitute a form of governance able to cope with the conditions and exigencies of the knowledge society" [13]. The author also associates the concept with "redesigning formal dem-ocratic governance" while maintaining the historically developed democratic prin-ciples and a free market economy. However, the definition of smart city governance is fragmented [14] and the many cities having smart city label often lack on a com-prehensive understanding about the nature of governance required in purview of digital revolution [15].

However, N.V. Lopes considers smart governance as a key factor for the imple-mentation of smart cities and achievement of its purposes by applying the appropri-ated policies. He maintains that the diversity of a city context, challenges, risks, and goals are unique factors in each city, and these factors require localized governance model that can enable and potentiate the creativity and innovation in the implemen-tation of smart cities [7].

In order to capture the dimensions of smart city governance, first literature has been retrieved from Web of Science and Google Scholar, with key words "Smart governance" and "Smart City Governance." However, the search also indicated two full-length sets of literature reviews on the subject: one by A. Meijer and M.P. Bolivar, 2015, titled "Governing the smart city: a review of the literature on smart urban governance" [14], and second by S. Praharaj et al. titled "Towards the Right model of Smart City Governance in India" [16]. Most of the research articles related to smart city governance have already been discussed in these reviews; therefore some of the findings on the typology of smart city governance have been adopted from these studies. Since the key aspect of this research is to suggest appropriate gover-nance model for smart cities in Pakistan, we could not find any suitable document on the topic. Moreover, no focused literature is available even on cities' governance. Available literature in Pakistan context generally covered "urban development issues," local governance issues in the context of overall decentralization, and a few policy briefs on the governance of specific cities like Lahore and Karachi. Therefore, effort has been made to draw a picture of current urban governance in Pakistan collecting scattered information from available sources coupled with public-sector experience of the co-author in Pakistan. In addition, the research on smart cities available in the repository of United Nation University in the Unit on Policy Driven e-Governance [17] has been referred as a guiding tool to examine the best practices and required components for smart city governance.

Detailed review of the literature on smart cities conducted by A. Meijer and M.P. Bolivar in 2015 [14], and lately by S. Praharaj et al. in 2018 [16], identifies four major types of smart city governance, varying in the degree of institutional transformation necessary to implement different types of smart cities, starting from least transformative type, where there is a concept of maintaining the existing governance structure for making the policy choices for an effective and efficient implementation of smart city initiatives [18]. Within this type of governance, government approves the development of a smart city and prioritizes some areas of action [19] and merely promotes smart city initiative [20] without transforming the existing structure. Generally, cities with such policies aim at adopting the "smart" label [15]. Such cities are often backed by global tech giants, create dazzling websites, and use glamorous social media contents to attract global attention and investment [16]. India's Gujarat International Finance Tech-City (GIFT smart city) and Canada's Edmonton are quoted as classic examples of this type where primary emphasis is on business environment and business-led economy [16].

Second type of conceptualization of smart governance focuses on smart decision-making through collecting real-time data to understand and solve cities' real-life challenges [16]. The model is closer to United Nations framework on good governance, i.e., "the process of decision-making and the process by which decisions are made" [21]. Sensor and network technologies are the pivot of this conceptualization. Walravens is of the view that decision-making can become innovative by using networks of technologies [22] whereas Schuurman et al. define smart governance as the process of collecting all kinds of data concerning public management by sensor networks [23]. Spanish smart city is considered as an ideal example of this model. The city [24] has dense concentration of installed sensors around the city's streets and a robust monitoring system is exploiting the Internet of Things (IoT) to unite all the information coming from sensors [16].

The third level of conceptualization of smart city governance entails a higher level of transformation. It is all about "Smart Administration" [25], i.e., restructuring and integrating internal government system, through electronic governance tools, supported by advanced digital technologies, to integrate internal government system. Batty et al. highlight that "smart governance is a much stronger intelligence function for coordinating the many different components that comprise the smart city. It is a structure that brings together traditional functions of government and business" [18]. Smart administration in this model breaks silos within the government departments by "interconnecting institutions, policies, information's and physical infrastructure to better service citizens and local communities" [16]. Smart administration model can be seen at its best in Singapore, having IoT foundational standards for information and service interoperability across infrastructure sectors. However, researchers also point out that manageable urban scale and the absence of overlapping state, local, and federal bureaucracies are the main reasons for success of the model in Singapore [16].

The fourth conceptualization of smart city governance focuses on urban collaboration as a major consideration. This type of governance involves a high-level transformation, as it requires integration of internal government structure, as well as partnership building with external organizations [14]. Nam and Pardo believe that smart governance primarily works through collaboration across government, industry, academia, nongovernmental organizations (NGOs), and people. This model of urban governance relies on the collective intelligence and creativity of the city dwellers [26]. Amsterdam Smart City [27] can be termed as the best example of this model which practices a unique partnership among various actors, i.e., the city government, businesses, research institutions, start-ups and innovators, investors, and common citizens. This model of smart governance may be considered the superlative type and very different from earlier conceptualizations that put the government at center stage and not people. All four types have been graphically depicted in Fig. 2.1, adopted from S. Praharaj et al. [16].

Study of different conceptualization provides a fair idea of the options and choices for the countries that have not started the journey of smart city and intend to leapfrog by developing a model suitable to their sociopolitical context. However, the choice for newcomers is still difficult due to certain complexities and confusions surrounding smart city governance. A. Meijer and M.P. Bolivar after extensive literature review of smart city governance hint that "the politics of smart cities have so far barely been analyzed" rendering the smart city an issue of puzzling nature. They indicate the following domains of confusion:

1. First domain of confusion is about the nature of smart city, i.e., technical or social. Some researchers have a technical focus while others emphasize the level of education of city inhabitants, whereas some combine these two perspectives in a socio-technical perspective on smart cities.
2. The second domain of confusion is whether smart city governance is mere "governance of a smart city" (first conceptualization) or an innovative way of decision-making, innovative administration, or even innovative forms of collaboration.
3. The third domain of confusion relates to the legitimacy claims of smart city governance. Some researchers consider that a city has a smart governance when it is sustainable or when citizens participate actively in governance. However a few academic publications also point out economic gains as legitimacy of smart city governance [14].

Another point which closely relates to the issue of politics and governance of smart cities is the role of local government structure in urban governance. City mayors have

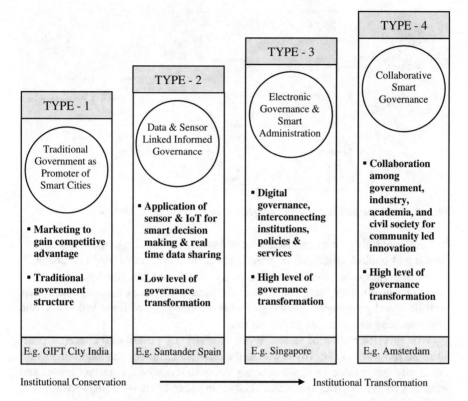

Fig. 2.1 Conceptualizations of smart city governance, adopted from S. Praharaj et al. [16]

a very important role in a good city governance [28]. They ensure the citizens' services, public good, and citizens' participation in local life [29]. Good governance for city dwellers entails five key aspects: (1) elected city governments; (2) city governments having capacity, power, and resources to act; (3) formal and informal avenues for civil society to influence and hold government accountable; (4) citizens' participation, particularly poor strata of urban population; and (5) rule of law not too biased against low-income groups [28]. This fairly justifies the role of local governments/city mayors in smart cities' governance.

In addition to the literature review conducted in two studies mentioned above, UNU-EGOV Reconnaissance Study on Smart Sustainable Cities [30] offers in-depth analyses of various attributes of smart cities on the basis of content analysis of 113 papers. The study looks at the governance attribute as part of smart city transformation which represents "how the Smart City government operates, how it manages public funds, how it delivers public infrastructure and services, how it supports sustainable city development, and how it engages its citizens in decision-making processes." As reflected in Fig. 2.2, governance and service delivery have been attributed to five smart principles, i.e., effectiveness, efficiency, transparency, collaboration, and openness. Functions and operations of smart city governance can

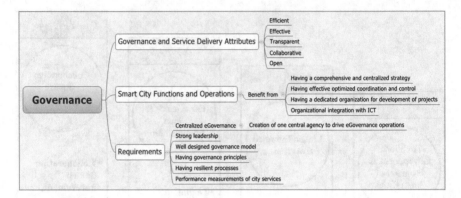

Fig. 2.2 Smart city governance attributes, adopted from UNU-EGOV, 2015 [30]

be beneficial if there is a comprehensive and centralized strategy, optimized coordination and control, dedicated organization for development of projects, and ICT-based organizational integration. The study also points out the requirements of smart city governance, which include centralized e-governance, strong leadership, well-designed governance model, governance principles, resilient processes, and performance measurements of city services.

With this description of various types of smart city governance concepts, practices, and confusions [14, 16] as well as the different attributes and requirements for smart city governance [30], we now are going to examine the current model and practices of urban governance in Pakistan, in order to frame a high-level model applicable to Pakistan context.

3   Cities' Governance in Pakistan

In Pakistan, the concept of urban governance led by elected city mayors is hardly visible. "The authority of the local government, the level of government closest to people, is restricted and overlaps with that of the provincial departments and local authorities" [31]. There is a long history of local government legislation, but city-level governments (municipal corporation/metropolitan corporation) are weak, with an institutional framework comprising general-purpose and single-function agencies operating at multiple tiers of government [32]. In fact, local governments are subservient to provincial governments. All four Pakistan provinces have their own local government Acts and a system of elected bodies also exist in urban as well as rural areas but "none of these Acts devolves sufficient functions and powers to the local governments, and all four provincial governments have retained the authority to suspend or remove the heads of an elected local government" [33]. Moreover, the functioning of the Local Government Fund is managed by the Finance Department and Finance Minister of the province [33]. Secondly, there is a challenge of fragmented governance. A number of organizations responsible for city planning, provision of the services, and development of

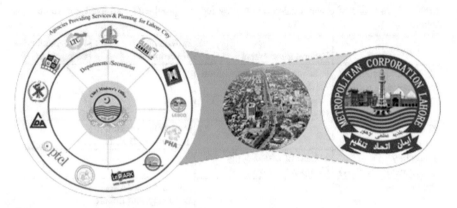

Fig. 2.3 Governance and service delivery in cities; example of Lahore, Pakistan

infrastructure are working in silos mostly under administrative control of provincial government, with overlapping functions and poor collaboration.

In order to elaborate the governance and service delivery mechanism example of Lahore, second largest city of Pakistan, has been depicted in Fig. 2.3. On the left side we have indicated various agencies actively engaged in service delivery (municipal services, ICTs, transport, development, communication, etc.) which are overwhelmingly under administrative control of provincial government departments reporting to the Chief Minister. On the right side there is local government (Lahore Metropolitan Authority) with nominal role in the city's governance and service delivery. The situation is more or less the same in all the other cities in Pakistan.

There are bureaucratic arrangements for coordination and administration of the cities headed by Deputy Commissioners at district level and Assistant Commissioners at Sub-District (tehsil) level. With fragmented and disintegrated governance "traditional establishments lack the essential technical expertise for current urban planning, have limited capacity and deficit of resources to deliver urban governance in an efficient manner" [4]. Therefore, for every significant innovative intervention federal and provincial governments have to make special arrangements, i.e., project management units, authorities, and companies. This is again evident from the example of Lahore city as it can be seen in Table 2.1, where a number of companies (i.e., put here the companies), authorities (i.e., put here the authorities), and units are working beside traditional service delivery agencies (i.e., put here the agencies). The majority of these agencies either are province wide or report to provincial departments.

4 Proposed Model

The Pakistan scenario described and discussed above shows how in Pakistan urban services are disintegrated and there is lack of a governance model ideal for smart cities. However, the present institutional and governance setup developed along the last 70 years cannot be undone and overcome overnight. Therefore, a balanced

Table 2.1 Example of urban service delivery by federal, provincial, and local governments in Lahore

	Federal government	Provincial government	Metropolitan corporation
Telecommunication infrastructure	Pakistan Telecommunication Authority and Pakistan Telecommunication Company		
IT and e-government services		Punjab Information Technology Board	
Electricity (infrastructure/supply)	Lahore Electric Supply Company		
Public transport		Punjab Mass Transit Authority, and Lahore Transport company	
Solid waste management			Lahore Waste Management Company
Emergency services		Rescue 1122	
Infrastructure development/city planning		Lahore Development Authority	
Housing and town planning		Punjab Housing and Town Planning Agency	
Parking			Lahore Parking Company
Water supply, sewerage, and drainage		Water Supply and Sanitation Agency (WASA) Lahore	
Parks and horticulture		Parks and Horticulture Authority, Lahore	
Tourism		Tourism Development Corporation of Punjab	
Environment		Environmental Protection Agency Punjab	

model towards smart governance is being proposed in this research work. Figure 2.4 shows the proposed model which is neither so ambitious to be impossible of being implemented nor so slack to compromise the overall principles of a smart city governance. In addition, the model is based on the principle of strong leadership while empowering local government with city mayors taking the responsibility of making their city smart. At the apex of smart city governance, there is a Smart City Steering Committee on each province.

Since local government is a devolved subject after 18th constitutional amendment in Pakistan and each province has its own local government legislation province-wise

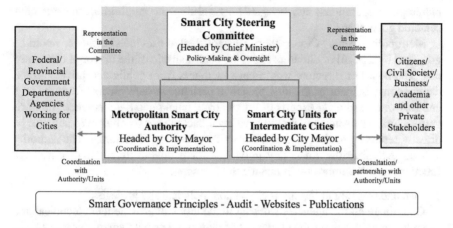

Fig. 2.4 Proposed smart city governance model for Pakistan

committees would be better positioned to develop a Smart City Vision/Mission/Plan for their respective provinces. The committees at federal level (for Smart City Islamabad only) would be headed by Prime Minister, whereas in provinces Chief Ministers would head the committee. Members of the committees will include mayors of respective smart cities, ministers of key ministries/departments (e.g., finance, planning, housing/urban development, IT), head of Metropolitan Smart City Authority, heads of smart city units, heads of major departments/autonomous bodies working in the cities, head of agencies implementing e-governance, and representatives of business, academia, and civil society. The committees will have the responsibility of leading the initiative, developing the vision and strategy, arranging resources, taking decisions for integration and alignment of ongoing development initiative with smart city plan where required, facilitating smart city legislation, facilitating the implementation tier (smart city authority), and monitoring and evaluating the execution of smart city initiatives.

Below the apex committees, a strong institutional arrangement is required to transform the smart city vision and strategy into practice through action plans, programs, and projects. For this purpose, two different models are being proposed: (1) metropolitan cities (provincial capitals) and (2) intermediate/small cities. In metropolitan cities (1) Metropolitan Smart City Authorities may be established as an autonomous specialized agency having corporate structure and manned by the experts of various smart city disciplines, whereas in intermediate/small cities (2) smart city units are proposed for each smart city within existing governance setup of the cities under the leadership of the mayor. Both the agencies would report to Smart City Steering Committee and work to implement smart city vision in accordance to the strategy decided at the apex level. These two slightly different models aim at (1) creating a balance between authorities of provincial governments and local governments; (2) availing the strong leadership of Prime Minister or Chief Minister for smart city transformation; (3) promoting the sense of an ideal city level of governance and service delivery among city mayors and city managers at metropolitan or municipal corporations; and (4) gradually shifting the equilibrium towards city-centered governance.

Metropolitan Smart City Authority, established at provincial capitals, would be governed by a board of directors under the chairmanship of the mayor of concerned city. The board of directors would have members from public and private sector, deputy commissioner of the city, proactive parliamentarians elected from the city, and renowned technical experts of urban development and e-governance from private sectors. The Authority would have legal backing for its objectives and will be directly reporting to the respective Smart City Steering Committee. The authority will contextualize the smart city initiatives for people-centered smart cities; the functions of this authority will include the folowing:

1. Plan and implement the smart city vision, policies, and projects.
2. Coordinate all the agencies working in the city and provide them technical and strategic guidance for the alignment of their projects with smart city vision.
3. Collaborate with all the stakeholders and citizens for a genuine engagement of citizens and to win their trust in the smart revolution of the city.
4. Play advisory functions for supporting the smart city units in intermediate and small cities. The authority should help these units in building their capacity as well as empowering them to steer independently.
5. Closely work with planning and development department of the provincial government, participating in the meetings of development forums and approving bodies.

For the smart cities of intermediate and small size smart city units are proposed within the existing administrative setup. The units would also be fully equipped to implement the interventions. However, these would have comparatively lean structure and would work within existing city governance setup under the mayor. The mayor of the city and head of the unit may represent the city in Provincial Steering Committee.

Overarching this two-tier governance model are the principles of smart city, i.e., effectiveness, efficiency, transparency, collaboration and openness, accountability and pluralism, and e-governance and e-government [7]. All the agencies involved in smart city execution would be subject to financial and performance audit per each smart city plan.

5 Conclusion

The study concludes that governance is the most critical component for a successful smart city transformation. Without a correct understanding of the meaning of a smart sustainable city the transformation which supposedly intends to be smart can be a fallacy and may even worsen the city situation. Therefore, a comprehensive understanding of all underlying smart city concepts and methodologies for the correct smart city transformation is required at the very beginning. This research work makes a thorough literature review on existing smart governance scientific papers as well as on policy documents about this topic, and then analyzes the particular case of Pakistan governance structure and conditions to finalize based on the previous two steps with the proposal of a suitable smart governance model for Pakistan context. The proposed two-tier smart governance model is flexible enough to be easily adjusted to Pakistan actual governance structure and to be gradually enhanced towards the ideal smart governance model for smart cities. The smart governance model is grounded in principles of what should be a good governance such as efficiency, transparency, collaboration and openness, accountability, and pluralism and the use of e-governance and e-government as an enabler and facilitator of those principles.

References

1. Pakistan Economic Survey: 2016–17, Economic Adviser's Wing, Finance Division, Government of Pakistan, www.finance.gov.pk. Accessed 10 April 2018
2. S. Nabi, Urban development in Punjab: a political economy analysis—policy brief, Consortium for Development Policy Research, http://cdpr.org.pk. Accessed 10 April 2018
3. Michael Kugelman, Understanding Pakistan's unstoppable urbanization, 2014, https://www.wilsoncenter.org. Accessed 11 Mar 2018
4. N. Jabeen, U.-e. Farwa, M. Jadoon, Urbanization in Pakistan: a governance perspective, 2017. J. Res. Soc. Pak. **54**(1) (2017)
5. Hina Saikh, Aijaz Nabi, 2017, The six biggest challenges facing Pakistan's urban future, International Growth Centre, www.theigc.org. Accessed 10 April 2018
6. Pakistan—Vision 2025, Planning Commission, Ministry of Finance, Government of Pakistan, http://pc.gov.pk/web/vision. Accessed 9 March 2018
7. N.V. Lopes, Smart governance: a key factor for smart cities implementation, in: *2017 IEEE International Conference on Smart Grid and Smart Cities (ICSGSC)*, Singapore (2017), pp. 277–282. doi: https://doi.org/10.1109/ICSGSC.2017.8038591
8. S. Joshi, S. Saxena, T. Godbole, Developing smart cities: an integrated framework. in: *6th International Conference on Advances on Computing & Communications, ICACC 2016*, 6–8 Sep 2016, Cochin, India (Elsevier B.V), http://creativecommons.org/licenses/by-nc-nd/4.0/
9. R. Giffinger, C. Fertner, H. Kramar, R. Kalasek, N. Pichler-Milanovi, E. Meijers, *Smart Cities: Ranking of European Medium-Sized Cities* (Centre of Regional Science (SRF), Vienna University of Technology, Vienna, Austria, 2007). Available from http://www.smartcities.eu/download/smart_cities_final_report.pdf.
10. N. Odendaal, Information and communication technology and local governance: Understanding the difference between cities in developed and emerging economies. Comput. Environ. Urban. Syst. **27**(6), 585–607 (2003)
11. R.M. Gisselquist, D. Resnick, Aiding government effectiveness in developing countries. Wind Energ. **34**, 141–148 (2014). https://doi.org/10.1002/pad.1694
12. A. Meijer, Smart city governance: a local emergent perspective, in *Smarter as the New Urban Agenda, vol 11*, ed. by J. Gil-Garcia, T. Pardo, T. Nam, (Public Administration and Information Technology, Springer, Cham, 2016)
13. H. Willke, *Smart Governance: Governing the Global Knowledge Society* (Campus Verlag, New York, NY, 2007)
14. A. Meijer, M.P. Bolivar, Governing the smart city: a review of the literature on smart urban governance. Int. Rev. Adm. Sci. **82**(2), 392–408 (2015)
15. R.G. Hollands, Will the real smart city please stand up? City **12**(3), 303–320 (2008). https://doi.org/10.1080/13604810802479126

16. S. Praharaj, J.H. Han, S. Hawken, Towards the right model of smart city governance in India, 2018. Int. J. Sust. Dev. Planning **13**(2), 171–186 (2018)
17. https://egov.unu.edu/ (accessed 10 April 2018)
18. M. Batty, K.W. Axhausen, F. Giannotti, A. Pozdnoukhov, A. Bazzani, M. Wachowicz, G. Ouzounis, Y. Portugali, Smart cities of the future. Eur. Phys. J. **214**, 481–518 (2012)
19. A. Alkandari, M. Alnasheet, I.F.T. Alshekhly, Smart cities: survey. J. Adv. Comput. Sci. Technol. Res. **2**(2), 79–90 (2012)
20. T. Nam, Modeling municipal service integration: a comparative case study of New York and Philadelphia 311 systems, Dissertation, University at Albany, State University of New York (2012)
21. UNESCAP, United Nations, ESCAP, Available at: http://www.unescap.org/resources/ what-good-governance. Accessed 16 Jan 2017
22. N. Walravens, Mobile business and the smart city: developing a business model framework to include public design parameters for mobile city services. J. Theor. Appl. Electron. Commer. Res. **7**(3), 121–135 (2012)
23. D. Schuurman, B. Baccarne, L. De Marez, P. Mechant, Smart ideas for smart cities: Investigating crowdsourcing for generating and selecting ideas for ICT innovation in a city context. J. Theor. Appl. Electron. Commer. Res. **7**(3), 49–62 (2012)
24. http://www.smartsantander.eu/. Accessed 10 April 2018
25. R. Gil-Garcia, Enacting electronic government success: an integrative study of government-wide websites, in *Organizational Capabilities, and Institutions*, ed. by R. Gil-Garcia, (Springer, New York, 2012)
26. T. Nam, T.A. Pardo, Smart city as urban innovation: focusing on management, policy, and context, in: *Proceedings of the 5th International Conference on Theory and Practice of Electronic Governance*. pp. 185–194 (2011)
27. https://amsterdamsmartcity.com
28. Environment & Urbanization Brief-18, What Role for Mayors in Good City Governance? 2009. International Institute for Environment and Development (IIED). https://www.iied.org/human/eandu/eandu_briefs.html
29. M. Diop, The role and place of mayors in the process of decentralization and municipal management in Senegal, in *Decentralization and the Politics of Urban Development in West Africa, Comparative Urban Studies Project*, ed. by D. Eyoh, R. Stren, (Woodrow Wilson International Center for Scholars, Washington DC, 2007), pp. 197–208
30. E. Estevez, N. Lopes, T. Janowski, Smart Cities for Sustainable Development Reconnaissance Study, UNU-EGOV (2015)
31. Khalida Ahson, Governance and management in Lahore, Centre for Public Policy and Governance Forman Christian College (A Chartered University) Lahore (2015)
32. A. Khan, Smart cities and infrastructure: challenges, issues and initiatives in Punjab Pakistan, 2016. https://www.researchgate.net/publication/301953878
33. S. Shafqat, Local government act 2013 and province-local government relations. Dev. Advocate Pak. **1**, 4–9 (2014)

Chapter 3
Building a Framework for Smart Cities: Strategy Development

Aroua Taamallah, Maha Khemaja, and Sami Faiz

1 Introduction

Compared to rural areas, cities play an important role in human life: they facilitate people's life and make it more flexible. Therefore, cities are attracting people to work and live there. This leads to exponential augmentation of the habitants of cities that will have reached 60% of the globe habitants by 2030 [1]. This exponential augmentation raises serious social, environmental, and economical challenges.

Decision makers call for smart city paradigm to mitigate these challenges and many others. They make use of innovative strategies to transform cities to smart ones. However, thinking and developing strategies is a repetitive and difficult process. It requires many efforts of stakeholders coming from various backgrounds, and data collections from various sources. Stakeholders have many concerns and points of view and use sources for city knowledge collection and strategy formulation. The research questions that we attempt to answer through this chapter are the following:

1. What are the main steps to carry on for successful development of smart city strategies?
2. How to formally represent knowledge coming from various sources and stakeholders and how to formally design strategies for their manipulation, adaptation, and simulation?

A. Taamallah (✉)
ISITCom, GP1 Hammam Sousse, Sousse, Tunisia

M. Khemaja
ISSATS, Taffala, Sousse, Tunisia

S. Faiz
ISAMM, Tunis, Tunisia
e-mail: sami.faiz@insat.rnu.tn

© Springer Nature Switzerland AG 2020
N. V. M. Lopes (ed.), *Smart Governance for Cities: Perspectives and Experiences*,
EAI/Springer Innovations in Communication and Computing,
https://doi.org/10.1007/978-3-030-22070-9_3

3. How to provide a common space to stakeholders for strategy development without considering time and space constraints?

To answer to the first question, we explore the existing knowledge about smart city strategies to identify steps of smart city strategy development. We consider the existing cases of smart cities as a useful source of information that we explored. We also validate the identified steps using examples from the chosen cases.

To answer the second question, we propose an ontology and a set of ontology design patterns for the representation of knowledge coming from various sources and stakeholders and to formally design strategies. The use of ODPs and ontologies provides a common language understandable by machine and human for an easy design, representation, and simulation of strategies.

To answer the third question, we provide a Web-based platform that can be ubiquitously used for the development of strategies. The Web-based platform offers a set of tools and services that help stakeholders during such a difficult process of strategy development.

The purpose is thus to highlight how to design, implement, and evaluate strategies for an effective and efficient development of a smart city. The proposal is therefore a framework that assists stakeholders during the strategy development. This aims to reduce deployed efforts and to achieve better results during the development process.

The remainder of the chapter is therefore structured as follows:

Section 3.2 includes an identification of the development process of smart city strategy; Sect. 3.3 includes a set of ontological models for strategies design; Sect. 3.4 describes the design and implementation of the strategy development platform; Sect. 3.5 proposes a use case for strategy design in educational domain and incorporates an evaluation of the proposed framework. Section 6 concludes and gives perspectives of the current work.

2 Identification of the Development Process of Smart City Strategy

2.1 Research Method Process Definition

To identify and provide the process behind successful strategy development, we attempted to define a methodological approach through a list of steps for the knowledge question that we attempted to answer which was to define the steps for successful development of smart city strategies. A survey of works related to strategy definitions, classifications, and frameworks is firstly done for a deeper problem investigation.

Secondly, a large volume of data is collected and knowledge about the studied cases is extracted. The cases of smart cities are selected; the smart cities are different; and they have different contexts (social, economic, and geographic). Thirdly, an analysis of existing strategies and projects at different stages was conducted according to the following criteria: problem identification, vision, mission and values definition, goal definition, objective definition, strategy definition, project definition, planning and implementation, evaluation, and decision-making. Fourthly, the analysis of existing initiatives allows us to identify the steps of smart city strategy development. Finally, the development process is validated through examples of selected smart cities.

2.2 Research Method Process Execution

2.2.1 Step 1: Related Work and Problem Investigation

In this section, we describe works related to strategy definitions, classifications, and frameworks for smart cities. These elements form the state of the art of strategies in smart cities and are essential for problem investigation.

Actually, there is no clear definition of a smart city strategy, yet a very few works try to give it a definition. The authors of [2] mention that a smart city strategy is "needed to overcome the existing gap for the actual delivery of smart services to the city's community and to achieve the smart city mission." Letaifa [3] considers a strategy as a roadmap for stakeholders to solve the city problems. She defines a strategy as "designing and steering a common vision of the city." Moreover, even though many works tried to propose taxonomy or strategy classifications, these taxonomies are quite different and do not tackle the same perspective. Anthopoulos and Blanas [4] distinguish between government strategy and domain strategy. They define government strategies as "tactic documents with which countries evolve or align to common supranational visions." Compared to government strategies, domain strategies are related to more centralized visions and are designed to evolve city domains related to a specific market or people (transportation, healthcare, economy, etc.).

Angelidou [5] uses a spatial perspective of smart city strategies and classifies those strategies into categories according to the development level of a city, the level of the strategy, the type of the infrastructure, and the application domain.

In strategy development process, elements such as vision, mission, success factors, and stakeholders are defined [4]. Contemporary governance emphasizes on engagement of citizens, institutions, researchers, governments, and other stakeholders in decision-making and planning in general [6]. Enhanced by information and communication technologies (ICT), a new terminology appeared such as e-participatory planning [7] and e-strategies [4].

Proposed frameworks are all oriented to the design and development of smart city strategy. Some of the proposed frameworks include an analysis of existing smart cities [8]; for example, they propose a framework composed of initiatives, policies domains, outcomes, enablers, critical success factors, and challenges. Others propose recommendations for a strong smart city strategy. Recommendations are the definition of a clear vision of the city, focusing on human capital instead of technology and on a specific topic that projects' developers should work under, a formal definition of a strategy, and the use of a smart city strategy framework that integrates resources and stakeholders [9]. Letaifa [3] introduces steps for strategizing a smart city through a model called SMART. The first step consists of the definition of a strategy. The second step is called multidisciplinary and includes resources and stakeholders coming from the different disciplines that are required for a successful transformation of the city. Appropriation is the third step where actors issued from different backgrounds collaborate to develop projects. The fourth step is a roadmap that consists of identifying an action plan for definition of projects. The final step consists of implementing the projects by providing services to end users.

The disadvantage of these works is that they do not consider a formal definition of strategies that is considered important for the proposition of effective strategies [9]. Letaifa [3] introduces the steps to follow during smart city strategy development.

None of the works existing in literature consider identification of the development process of strategies from smart cities' existing cases, formal modeling, and structuring of data coming from various sources and stakeholders for strategy formulation especially those used for formal design of strategies and providing a Web-based platform for stakeholders as a space of interaction and strategy delivery.

The contribution of this chapter includes therefore (1) strategy development process, (2) ontological models used during these steps for knowledge structuring and strategy modeling, (3) a Web-based platform to facilitate the development process of smart city strategies.

2.2.2   Step 2: Data Collection and Knowledge Extraction

The process of knowledge extraction is based on automatic annotation and Ontology Based Information Extraction (OBIE). It is composed of two steps: ontology construction and ontology population (Fig. 3.1). This process is out of the scope of this chapter and described in detail in Taamallah et al. [10].

Data are collected from various sources such as web sites, research papers, strategies, and strategic plan documents about the smart cities of Vienna, Amsterdam, Barcelona, New Castle, Edmonton, and London. The collected data is extracted from documents that are most recently published (2014–2017). Publications related to the selected smart cities are searched using key words' combination including the name of the smart city and the word strategy or strategic plan (e.g., Vienna Smart City Strategy or Vienna Smart City Strategic Plan). We obtain in total 870 documents classified into two categories (see Table 3.1): the first category includes documents aligned to the topic of the present wok and the second one includes irrelevant documents. For the city of Vienna for example, 105 documents deal with the smart city strategy, strategic plans, projects, and initiatives and 20 others out of 125 are irrelevant.

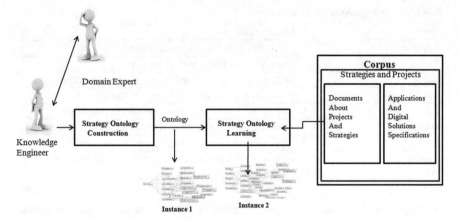

Fig. 3.1 Knowledge extraction process

Table 3.1 Corpus construction

City	Relevance		
	Yes	No	Total
Vienna	105	20	125
Amsterdam	149	13	162
Barcelona	99	30	129
New Castle	106	17	123
Edmonton	165	22	187
London	130	14	144
Total	754	116	870

2.2.3 Step 3: Case Description

A summary of knowledge extracted from smart cities of Vienna, Amsterdam, Barcelona, New Castle, Edmonton, and London is described in this section.

The vocabulary used for case description is illustrated in Table 3.2.

Smart city descriptions are structured in a uniform manner: Firstly, we identify if smart cities are smart or not, and if they have ranks and awards or qualifications. Secondly, we identify if there exist problems that the city suffers from, and visions, goals, objectives, and strategies to transform the city to a smart one. Finally, we describe examples of projects implementing strategies.

Case 1: Vienna

Vienna is ranked as the first among 100 cities in quality of life (QoL), innovation, and information and communication technologies (ICT). The main problems of Vienna are increased CO_2 emission and excessive consumption of fossil fuels. The Vienna 2050 strategic plan [11] aims to reduce CO_2 emission by conserving resources and to offer a high QoL and social participation. The participants in the

Table 3.2 Description of used vocabulary

Term	Description
Problem	A challenge that the city suffers from
Vision	Describes the future state of the city
Value	Regroups cultures of participants in the strategy definition
Mission	Includes goals to achieve during the strategy definition
Goal	A goal is an intended achievement for the long term
Objective	An objective is an intended achievement for the short term
Strategy	A strategy is a plan intended to achieve an objective
Project	A project is an initiative or a program to implement the strategy

development of the strategic plan are citizens, enterprises, nonprofit institutions, and the public sector. "Vienna is a livable city to all (children, young people, elderly)" is the 2050 vision. The Smart City Wien Framework strategy integrates specialized strategies in transport, energy, and climate. Energy Strategy 2030, Renewable Energy Action Plan, and STEP 2025 are examples of projects to reduce gas emissions (80% by 2050), to preserve resources and use renewable energy resources. Climate Protection Program is an example of successful projects. It reduced gas emissions by 10% in 2010.

Case 2: Amsterdam

Amsterdam is one of the well-known smart cities around the world. In 2016, it received the Europe capital of innovation award since high innovative efforts have been done to offer efficient services and to improve citizens' quality of life. Amsterdam is characterized by a diversified population. It also suffers from problems such as weather extremity, energy waste, and its elderly population.

Amsterdam smart city program aims to reduce CO_2 emission and energy consumption through more than 70 projects. The strategy is achieved thanks to the collaboration of its founders the Amsterdam Innovation Motor (AIM) and the municipal administration. The engagement of citizens also participates in the success of many projects such as smart citizen kit, Vehicle2Grid, and city-Zen. To measure air quality, the Amsterdam's government affords the Smart Citizen Kit that aims to measure air quality. This kit measures humidity, noise pollution, light intensity, temperature, and CO_2 and NO_2 gazes. The citizens are involved in the process of measurement to be conscious of their energy consumption. They use a platform to publish collected data. The Vehicle2Grid is a project that integrates smart transportation and smart energy to create a clean environment. Solar energy is produced from solar panels and transferred to smart grids. Through the grids, electronic vehicles (EV)'s batteries are charged not only to drive the car but also to run house devices. The unused energy is returned to grids.

Case 3: Barcelona

Barcelona is considered as one of the top smart cities in the world. In 2014, it received the European Capital of Innovation "iCapital" prize. Its smart city strategy aims to improve the citizens' QoL. It also utilizes knowledge to foster economic progress and utilizes information and communication technologies (ICI) to improve urban functions. The strategy includes more than 100 projects. These projects aim to regenerate the city (e.g., Superblocks project) and to provide both physical and digital innovative solutions. Through the collaboration of Barcelona City Council and the Urban Ecology Agency, the Superblocks project aims to enhance the human dimension of the city of Barcelona by (1) providing more sustainable mobility, (2) fostering biodiversity and urban vegetation, and (3) including people in social life and governmental processes.

Barcelona Open Government is another project that integrates citizens in the governmental process. Thanks to ICT, citizens can access information easily, visit spaces for dialogue and cooperation, and collaborate with the City Council to create new policies. New Bus Network project provides a smart city's network that allows the traveler to move easily around the city and informs him or her about any improvements. Additionally, the network considers children and elderly. Thanks to smart cycling project, citizens have access to more than 400 bike stations.

Case 4: New Castle
New Castle vision is "open, collaborative and connected, livable and sustainable city for all" [12]. It includes principles that come out from the process of collaboration, connectivity, efficiency, openness, and people. The New Castle strategy is structured according to domains which are smart governance, smart environment, smart living, smart mobility, smart people, and smart economy. For smart people, for example, the objective is to invest in people and encourage them to use smart technologies. The strategies that aim to achieve this objective are (1) to provide an open access to city data to improve planning and economic development and to attract people and improve their life quality, (2) to ensure equitable and easy access to digital technologies, (3) to provide access to city digital platforms and support creative industries and broaden audiences, and (4) to enable communities' communications by building their capacity in understanding and using smart technologies. The planned projects are aligned to the strategies. For example, the city data project is a software that helps citizens to easily access city services and data.

Case 5: Edmonton
Edmonton's problems are rapidly changing community and surrounding areas. This phenomenon affects the economic, social, and political factors of the city. Edmonton smart city strategy addresses these problems and makes future opportunities through a creative community of government, academia, citizens, and industry. It includes three goals: resiliency, livability, and workability. For each of the goals, strategic dimensions are created. The vision behind resiliency theme is to obtain a well-planned and adaptive municipality and to offer adaptive processes for community future plans. Objectives are aligned to each of the strategic dimensions. After that, strategy initiatives are mapped to strategic dimensions. Concerning the resiliency with focus on citizen theme for example, the objectives are (1) to develop strategies that ensure citizens' engagement and address their needs and (2) to ensure access to services for all citizens. Edmonton Service Center, idea generator, and online

engagement are implemented projects to achieve the identified objectives. For example, Edmonton Service Center project offers services related to payments, permits, licenses, and more to citizens in one location.

Case 6: London
London is ranked as the third smart city in the globe. Many strategies are adopted to transform it to a smart one [13]. [14] is one of these. This plan is divided into four phases. The first phase ended in 2014 and aimed to solve problems related to homeless and housing. The intended visions are "All members of our community have access to housing that is safe, secure and suitable to their needs and ability to pay." And "A coordinated and integrated individual and family centered-housing stability approach that addresses, reduces and prevents homelessness in London." The two main goals are to enable homeless people to obtain housing and people at the risk of homelessness to remain housed. The goals are aligned to four objectives which are securing housing, achieving housing stability, emergency shelter division, and housing with support. The strengths of the project are that all people recognized as homeless obtain houses. The project continues to recognize more homeless people in the future phases of the strategy. The project will also consider collaboration with housing and community partners and addresses the unique needs of individuals and families.

2.2.4   Step 3: Discussion and Comparison

The steps are identified from analysis and observation of smart cities of Vienna, Amsterdam, Barcelona, New Castle, Edmonton, and London.

The identified steps are summarized and compared. The results of the comparison are illustrated in Fig. 3.2. All the selected cities follow the same process for strategy development. They identify the city problems; create a common vision of the city; identify the goals, objectives, and strategies; and implement and evaluate the strategies.

Sometimes the city problems are explicitly included in the strategy and are not clearly described from the beginning of the documentation (e.g., New Castle and Barcelona) (see Fig. 3.2). New Castle Strategy is characterized by a planning context. The planning context creates relationships with existing and upcoming strategies at a local, regional, or international level. This characteristic is not found in the other strategies. For the Vienna smart city, actors do not identify a vision for all the cities. For each of the identified goals, a vision is affected. All the cities clearly describe goals and objectives that strategies should achieve. They also describe in detail the planned projects. If the project is implemented, the obtained results are also described to evaluate it and then evaluate the success degree of the strategy. For the cities of Barcelona and Amsterdam, the strategy is holistic for all the cities and projects belonging to strategic directions are implemented to achieve the strategy. For the cities of Vienna, Edmonton, and New Castle, they identify sub-strategies for each of the previously defined strategic directions.

Steps	Vienna	Amsterdam	Barcelona	New Castle	Edmonton	London
Problem Identification	Δ	Δ	◇	◇	Δ	◇
Vision, mission and values definition	Δ	◇	Δ	Δ	Δ	Δ
Goals definition	Δ	Δ	Δ	Δ	Δ	Δ
Objectives definition	Δ	Δ	Δ	Δ	Δ	Δ
Strategy definition	Δ	Δ	Δ	Δ	Δ	Δ
Project definition, planning and implementation	Δ	Δ	Δ	Δ	Δ	Δ
Evaluation and decision making	Δ	Δ	Δ	Δ	Δ	Δ

◇ Explicitly defined, Δclearly defined, O non-defined

Fig. 3.2 Smart city comparison and step identification

2.2.5 Step 4: A Generalized Development Process of Smart City Strategies

What comes out is that all the previously discussed smart city cases follow the same process. This process explains how to construct effective and successful smart city strategies. It guides actors participating in a strategy design, implementation, and evaluation. This process is divided into four phases: pre-design phase, design phase, implementation phase, and evaluation phase. Figure 3.3 illustrates the strategy development process in a smart city.

The pre-design phase includes activities related to knowledge resource analysis and city challenge extraction which are useful for the next phase (design phase). It includes one step which is learning from smart city initiatives at an international level. As an example of approach for automatic learning from smart city initiatives an approach based on documents' automatic annotation and ontology learning techniques is proposed in Taamallah et al. [10].

The second phase is strategy design phase. It includes the following steps:

The second step includes collecting information about the city and identifying its problems. Stakeholders consult several literature resources such as statistics, strategy documents, and reports; cities' web sites (e.g., https://amsterdamsmart-city.com/); questionnaires; and experts' interview. They can also do researches themselves when the collected data is insufficient or inaccurate. Then, they extract problems. The third step comprises the vision identification where the stakeholders identify what the city will look like in the future. The vision should be clear, powerful, and specific to the city. The third step includes mission and value definition. The stakeholders define the strategic directions (mission) sharing various values such as collaboration and cooperation, and responsibility and innovation. In the fourth step, they identify the goals to achieve. The goals are determined with reference to identified problems by aligning each problem to each goal as

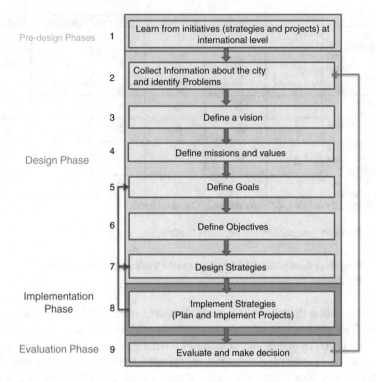

Pre-design Phases 1   Learn from initiatives (strategies and projects) at international level

2   Collect Information about the city and identify Problems

3   Define a vision

Design Phase 4   Define missions and values

5   Define Goals

6   Define Objectives

7   Design Strategies

Implementation Phase 8   Implement Strategies (Plan and Implement Projects)

Evaluation Phase 9   Evaluate and make decision

Fig. 3.3 Strategy development process

a desired solution to that problem. During the fifth step, objectives supporting each goal are determined. Conversely to goals which are intended achievements for a long term, objectives are intended achievements for a short term. The sixth step comprises strategy definition. The stakeholders define strategies that might achieve the identified objectives. Two categories of strategies may be defined as top-level strategies and sub-strategies. A top-level strategy describes a more abstract plan while sub-strategies (or domain strategies) are defined under strategic axes. During the seventh step, the defined strategies are implemented by means of scheduled projects (attributing financial and human resources). Projects are periodically evaluated during the implementation phase. The evaluators verify if it produces a positive impact and if the planned objectives are achieved. When the implementation is finished, the results are discussed to make decisions about the strategy, identify issues and opportunities, and define goals and challenges for future improvements.

2.2.6   Step 5: Validation of Smart Cities' Strategy Development Process

The development process of smart city strategies is validated using examples from the selected smart cities. For each of the identified steps of the development process of smart city strategies, we extract instances that allow to validate the proposed process. Table 3.3 illustrates the validation.

3   Ontological Models for Knowledge Repreparation and Strategy Design

Ontological models described in this section have mainly two aims which are (1) data coming from various sources and from various stakeholders should be well structured, and (2) formal design of strategies and thus a machine- and human common-understandable language.

Ontological models (or Ontology Design Patterns (ODP)) are described by specifying the step of the development process of strategies where each model is used. Those models are then integrated in a Web-based platform to help stakeholders during strategy development.

The pre-design phase includes one step and does not require deposition of the ontological model. For that aim, an ontology is used in this phase. However, the design phase includes many steps and we use ODPs as modules. The use of ODPs instead of a holistic ontology facilitates their reusability, extensibility, and adaptability. ODPs can be used separately or together according to stakeholders' needs during the design process of smart city strategies for formal specification and structuring of strategies.

3.1   Step 1 and Strategy Ontology

Strategy Ontology is a domain ontology (Fig. 3.4) that aims to acquire expertise from existing projects and strategies. Automatic learning techniques are used to enrich strategy ontology with instances from real smart city documentation [10]. The resulting knowledge helps stakeholders to learn from experiences of smart cities.

Table 3.3 Validating the process with examples from selected smart cities

Steps	Vienna	Amsterdam	Barcelona	New Castle	Edmonton	London
Problem identification	Increased CO_2 emission Excessive consumption of fossil fuels	The most diversified population in Europe The residents are living longer Extremity of weather. Energy waste	Access to housing Energy waste	Difficulty in keeping its graduated younger generation Not enough high-paying jobs A need to provide better economic opportunities	Edmonton community and regions change rapidly	Housing needs and homelessness
Vision, mission, and value definition	Vienna is a livable city to all	Amsterdam 2040 vision for livable and sustainable city	By 2020, Barcelona is a more sustainable, smart city	Open, collaborative and connected, livable, and sustainable city for all	To obtain resilient, well-planned, and adaptive municipality. To offer adaptive processes for community future plans	Safe and secure access to housing Preventing and reducing homelessness in London
Goals definition	Conserve resources offer a high QoL and social participation	Reduce energy waste and CO_2 emission (by 40% in 2025) Promote sustainable economic growth	Renewable energy Rethink urban planning Enhance citizens' engagement	Stakeholders' collaboration Renewable and smart technology adoption Self-promotion of the city	Resiliency Livability Workability	Homeless people obtain housing People at risk of homelessness remain housed
Objectives definition	Reduce gas emission by 80% by 2050 50% of the energy consumption comes from renewable energy	Intensive use of public spaces Public-private partnerships High life quality Sustainable energy Green spaces and water	Improve the citizens' quality of life	To invest in people and encourage them to use smart technologies	Develop strategies that facilitate citizens' engagement Develop strategies that address citizens' needs Ensure access to services for all citizens	Secure housing Achieve housing stability Emergency shelter Diversion Housing with support

			Barcelona smart city strategy		Edmonton Smart City Strategy 2017	Homeless prevention and housing plan 2010–2014
Strategy definition	Strategy 1: Active participation of people in resource conservation Strategy 2: Intensive use of renewable energy	The strategy aims to accelerate energy and climate programs		Strategy 1: an open access to city data Strategy 2: an equitable and easy access to digital technologies		
Strategy implementation (examples of projects)	Climate protection program	Smart citizen kit	Smart cycling	City data project	Edmonton service center	London homeless prevention system
Evaluation and decision-making	Gas emission decrease by 10% in 2010	100 citizens participate by sending their energy consumption. Citizens become conscious of their consumption and participate to reduce it	Access to more than 400 bike stations Barcelona is the second smart city in transport	Access to city data and services for all	Services related to payments, permits, licenses, and more to citizens in one location	Homeless people obtain houses Collaboration with housing and community partners Address the unique needs of individuals and families

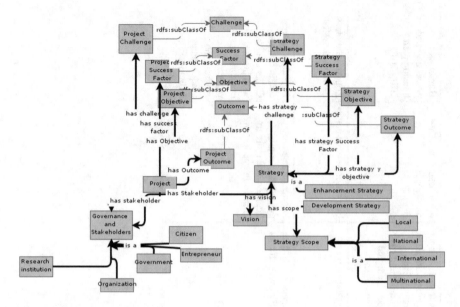

Fig. 3.4 Strategy ontology

3.2 Step 2 and Problem-Cause-Effect ODP

Stakeholders coming from various domains identify the city problems, then analyze, and model them using the problem-cause-effect ODP (see Fig. 3.5). The problem-cause-effect ODP is inspired from Problem Tree [15]. Problem Tree is used for planning projects. It represents problems, their causes, and effects.

The problem is firstly defined using a pattern named Problem ODP. Then, it is analyzed using problem-cause-effect ODP. Figure 3.6 contains its related axioms.

3.3 Step 3 and Problem-Solution-Vision ODP

When the stakeholders identify the city problems, they can reverse the negative statements of the problem into positive ones; this is what we call solution (e.g., transform the problem "polluted environment" into "a not polluted environment"). Then, they form an idea of the vision of the city. The problem-solution-vision pattern helps stakeholders to define the city vision (see Fig. 3.7).

Figure 3.8 illustrates axioms related to problem-solution-vision ODP.

Fig. 3.5 Problem-cause-effect ODP

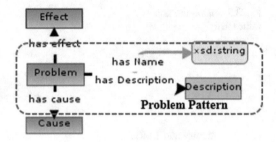

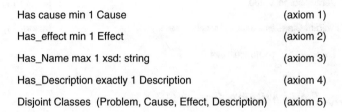

Has cause min 1 Cause (axiom 1)

Has_effect min 1 Effect (axiom 2)

Has_Name max 1 xsd: string (axiom 3)

Has_Description exactly 1 Description (axiom 4)

Disjoint Classes (Problem, Cause, Effect, Description) (axiom 5)

Fig. 3.6 Problem-cause-effect ODP axioms

Fig. 3.7 Problem-solution-vision ODP

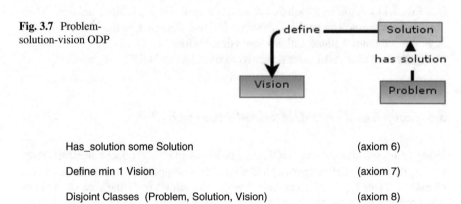

Has_solution some Solution (axiom 6)

Define min 1 Vision (axiom 7)

Disjoint Classes (Problem, Solution, Vision) (axiom 8)

Fig. 3.8 Problem-solution-vision ODP axioms

3.4 Step 4 and Vision-Mission-Value ODP

Using vision-mission-value ODP, stakeholders can model missions and values with reference to the defined vision (see Fig. 3.9). A well-defined and clear vision allows to identify at least one mission that the stakeholders should accomplish. The identified mission(s) allow to identify the participants' values. The classes vision, mission, and value are disjoint.

Figure 3.10 illustrates axioms related to vision-mission-value ODP.

Fig. 3.9 Vision-mission-
value ODP

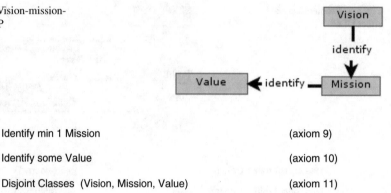

Identify min 1 Mission   (axiom 9)

Identify some Value   (axiom 10)

Disjoint Classes (Vision, Mission, Value)   (axiom 11)

Fig. 3.10 Vision-mission-value axioms

3.5   Step 5 and Mission-Goal ODP

The mission-goal ODP allows stakeholders to model goals included in a mission
(see Fig. 3.11). A mission includes at least one goal. Goal, mission, and description
are disjoint classes. The goal-ODP is intended to define each of the identified goals.
A goal has au max 1 name and has some descriptions.

Figure 3.12 illustrates axioms related to mission-goal ODP.

3.6   Step 6 and Goal-Objective-Strategy ODP

Using goal-objective-strategy ODP, stakeholders can easily model objective(s) supporting each of the defined goal(s) and strategies(s) supporting each of the identified
objective(s) (see Fig. 3.13). The objective-pattern allows to define each of the identified objectives where an objective has au max 1 name and has some descriptions.
In the goal-objective-strategy ODP, each goal is supported by some values of objectives, and each objective is supported by some strategies. Goal, objective, description, and strategy classes are disjoint.

Table 3.4 includes rules related to the goal-objective-strategy ODP.
Figure 3.14 illustrates axioms related to goal-objective-strategy ODP.

3.7   Step 7 and Strategy ODP

Strategy ODP allows stakeholders to formally design strategies (see Fig. 3.15). The
main concepts to define a strategy are name, strategy description, challenge, context, strategy success factors, and scope. Context concept mentions contextual

Fig. 3.11 Mission-goal ODP

Has_description some Description	(axiom 12)
Has_Name max 1 xsd: string	(axiom 13)
Include min 1 Goal	(axiom 14)
Disjoint Classes (Goal, Mission, Description)	(axiom 15)

Fig. 3.12 Mission-goal ODP axioms

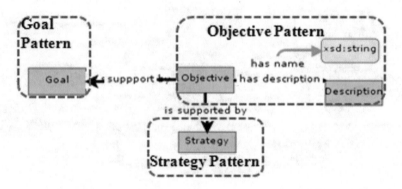

Fig. 3.13 Goal-objective-strategy ODP

information related to the strategy including city context and planning context while strategy success factors are influencers that affect the fulfillment of the strategy. The scope of the strategy can be local, national, and international. A strategy is implemented by one or more projects. Strategy, challenge, strategy success factor, context, scope, and project are disjoint classes.

Figure 3.16 contains the axioms related to strategy ODP.

Table 3.5 shows the strategy ODP rules.

Table 3.4 Goal-objective-strategy ODP rules

Rule	Explanation
Goal (? x) ^ Objective (? y) ^ Strategy (? z) ^is supported by (? x, ? y) ^ is supported by (? y, ? z) -> is supported by (? x, ? z)	If a goal is supported by an objective and an objective is supported by a strategy, then the goal is supported by the strategy.
Goal (? x) ^ Objective (? y) ^ Strategy (? z) ^ is supported by (? z, ? y) ^ support (? y,? x) -> is support by (? z, ? z)	If a strategy supports an objective and an objective supports a goal, then the strategy supports the goal.

Is_supported_by some Objective   (axiom 16)

Is_supported_by some Strategy   (axiom 17)

Has_name max 1 xsd: string   (axiom 18)

Has_description some Description   (axiom 19)

Disjoint Classes (Goal, Description, Objective, Strategy)   (axiom 20)

Fig. 3.14 Goal-objective-strategy ODP axioms

Fig. 3.15 Strategy ODP

3.8 Step 8 and Project ODP

The project ODP allows stakeholders to formally design projects related to each of the defined strategies (see Fig. 3.17).

A project's concepts are project name, project descriptions, project success factors, project objectives, project challenge, and planning. When we plan a project Time Pattern is used to indicate the beginning and the end or the duration of the scheduled actions. The classes project, project description, project objective, project challenge, planning, and project success factor are disjoint. Figure 3.18 contains the axioms related to strategy ODP.

Has_name max1 xsd: string   (axiom 21)

Has_description some Description   (axiom 22)

Has_challenge some Challenge   (axiom 23)

Has_scope some Scope   (axiom 24)

Has_success_factor some Strategy_Success_Factor   (axiom 25)

Is_implemented_by some project   (axiom 26)

Has_context some Context   (axiom 27)

Disjoint Classes (Strategy, Description, Scope, Challenge,   (axiom 28)
Context, StrategySuccess_Facto)

Fig. 3.16 Strategy ODP axioms

Table 3.5 Strategy ODP rules

Rule	Explanation
Existing City (? x) ^ Strategy (? y) ^ has Strategy (? x, ? y) -> Enhancement Strategy (? y)	If an existing city has a strategy then the strategy is an enhancement one.
New City (? x) ^ Strategy (? y) ^ has Strategy (? x, ? y) -> Development Strategy (? y)	If a new city has a strategy then the strategy is a development one.
Strategy (? x) ^ Country (? y) ^ is applied to (? x, ? y) -> has scope (? x, local)	If a strategy is applied to a country, the strategy scope is national.
Strategy (? x) ^ City (? y) ^ is applied to(? x, ? y) -> has scope (? x, local)	If a strategy is applied to a city, the strategy scope is local.

3.9   Summary of Correspondence Between Steps and Models

To resume, for each step of the development process corresponds an ontological model (or an ODP) that helps for knowledge structuring and for developing formal strategies. Figure 3.19 illustrates that correspondence.

4   Strategy Development Platform Design and Implementation

The strategy development platform is offered to help and assist stakeholders during the development of strategies. It provides a space for discussion, communication, knowledge sharing, and strategy formulation. Figure 3.20 illustrates the architecture of strategy development platform.

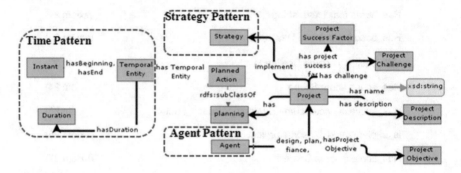

Fig. 3.17 Project ODP

Has_name max 1 xsd: string	(axiom 29)
Has_description some Project Description	(axiom 30)
Implement some Strategy	(axiom 31)
Has some planning	(axiom 32)
Has_project_success_factor some Project Success Factor	(axiom 33)

Fig. 3.18 Project ODP axioms

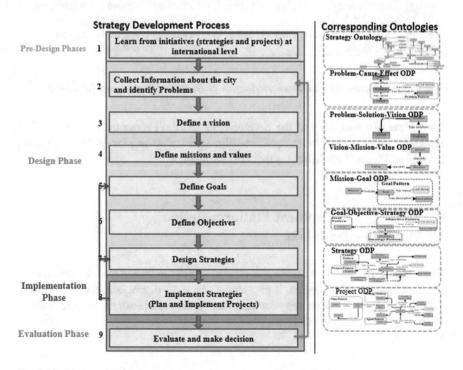

Fig. 3.19 Strategy development process and its corresponding ontologies

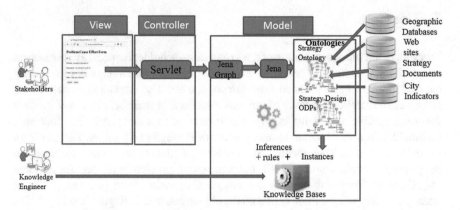

Fig. 3.20 Strategy development architecture

The strategy development platform is implemented using Spring Model-View-Controller (MVC) framework [16]. The latter separates between the models and their representation (i.e., the view).

The ontologies and ODPs are implemented in the OWL2 (Ontology Web Language version 2) language by making use of the Protégé ontology development environment. Apache Jena Engine is a reasoner that helps to do inferences and extract new knowledge. It is also used to store data in ODPs using Jena Graph.

The stakeholders use the strategy development platform as follows:

For the phase of pre-design, the ontology is already enriched with instances from smart city strategies and is provided to stakeholders as a Web service to acquire expertise from real initiatives. Stakeholders can also enrich the provided knowledge base with new initiatives using Web forms.

For the phase of design, stakeholders choose the design step that they will work on. Each stakeholder uses Web forms corresponding to that step to provide its opinions and ideas. The ideas are added to ODPs as RDF tuples using Jena Graph. The stakeholders share the ideas through a collaborative space (e.g., Google Docs). After discussions, the retained knowledge is stored in ODPs. The ODPs can then be queried and strategies' related knowledge is visualized, adapted, and shared. Other participants such as citizens and businesses can then access the platform to give their opinions about the strategies or to participate in their implementation.

For the implantation and evaluation phases, they are under development. We will provide a smart city simulator that allows to test and evaluate strategies before we implement them on cities. This will allow to gain money and time and to avoid risks and problems that can affect city and citizens.

5   Use Case

To validate the proposed framework, a group of stakeholders especially researchers
and decision makers have tested the proposed platform, for strategy design in educational domain in a Tunisian city named Sousse. The education improvement
increases the intellectual level of citizens, ameliorates their behavior, and renders
them responsible and active in city transformation to a smart one. For that aim,
stakeholders follow the development process of strategies and use the Web platform
to design strategies. They identify the problems of educational system especially in
the primary school. Figure 3.21 illustrates the use of problem Web form for problem
identification. They also identify the vision using vision form (see Fig. 3.22). In
addition to that, stakeholders define goals, objectives, and strategies (see Fig. 3.23).

SPARQL queries are used to visualize strategies' related knowledge. Figure 3.24
shows educational domain problems with their causes and effects. Figure 3.25
shows goals, objectives, and strategies.

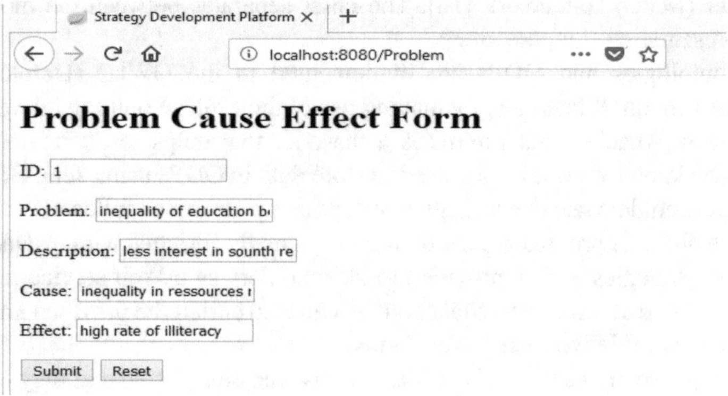

Fig. 3.21 Screenshot of problem identification

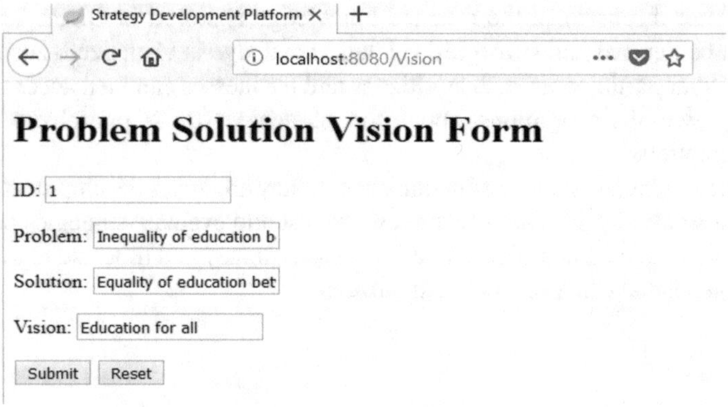

Fig. 3.22 Screenshot of vision identification

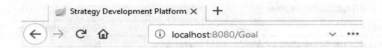

Fig. 3.23 Goal, objective, and strategy identification

```
PREFIX : <http://www.semanticweb.org/BlemTree#>
PREFIX rdf: <http://www.w3.org/1999/02/22-rdf-syntax-ns#>

SELECT ?problem ?cause ?effect
WHERE {
     ?problem :has_cause ?cause.
     ?problem :has_effect ?effect.

}
```

problem	cause	effect
:inequality_of_education_between_regions	:lack_of_schools_in_rural_areas	:students_coming_from_pover_families
:inequality_of_education_between_regions	:lack_of_schools_in_rural_areas	:high_rate_ol_illitracy
:Old_Curricula	:long_learning_hours	:irritated_students
:Old_Curricula	:long_learning_hours	:education_hate
:Old_Curricula	:long_learning_hours	:boredom_of_students
:bad_school_infrastructure	:schools_are_far_from_students_homes	:ruined_schools
:bad_school_infrastructure	:schools_are_far_from_students_homes	:private_tutoring
:bad_school_infrastructure	:schools_are_far_from_students_homes	:insufficient_schools_in_cities
:inequality_of_education_between_regions	:schools_are_far_from_students_homes	:students_coming_from_pover_families
:inequality_of_education_between_regions	:schools_are_far_from_students_homes	:high_rate_ol_illitracy
:gifted_students_with_difficulties_in_schools	:depression	:decrease_in_school_level
:gifted_students_with_difficulties_in_schools	:depression	:Aggressive_behavior
:Old_Curricula	:no_consideration_of_new_technologies	:irritated_students
:Old_Curricula	:no_consideration_of_new_technologies	:education_hate
:Old_Curricula	:no_consideration_of_new_technologies	:boredom_of_students
:bad_school_infrastructure	:less_maintenance	:ruined_schools
:bad_school_infrastructure	:less_maintenance	:private_tutoring
:bad_school_infrastructure	:less_maintenance	:insufficient_schools_in_cities

Fig. 3.24 SPARQL query showing educational domain problems

```
PREFIX strategy: <http://www.semanticweb.org/Strategy#>
PREFIX project: <http://www.semanticweb.org/project#>
PREFIX rdf: <http://www.w3.org/1999/02/22-rdf-syntax-ns#>

SELECT ?strategy  ?project ?planningContext
WHERE {  ?strategy strategy:hasplanningContext ?planningContext.
         ?strategy strategy:isImplementedBy ?project.

      }
```

strategy	project	planningContext
strategy:Educating_smart_technologies	strategy:Include_IoT_in_learning_scenarios	strategy:PreviousStrategy_Modernizing_Infrastructure
strategy:Educating_smart_technologies	strategy:Include_IoT_in_learning_scenarios	strategy:PreviousProject_Madrasaty

Fig. 3.25 SPARQL query illustrating designed strategies

Table 3.6 Platform evaluation

Question	Response yes/no/maybe
Do you find the use of strategy development platform useful?	80% yes 20% no
Do you spend less effort in strategy design using the platform?	75% yes 25% maybe
Does the use of ontologies provide a common definition and understanding of strategies	92% yes 8% maybe
Do you find the platform useful when you are not in the same place with other stakeholders?	99% yes 1% no (prefer real-time discussions)

To evaluate the platform, the stakeholders answer to some questions.

The preliminary results of the use of the platform show that the platform has been useful for stakeholders in the process of strategy development (see Table 3.6). It facilitates the process of strategy development and removes the time and place limits. The use of ontologies helps in formal design and representation of strategies and provides a common machine- and human-understandable language.

5.1   Conclusion

Smart city aims to mitigate current problems in cities. Stakeholders call for innovative strategies to transform cities to smart ones. The development process of strategies is hard, complex, and repetitive. Identification and implementation of such a process facilitate efforts performed by stakeholders. The development process of strategies is identified from knowledge about existing smart city initiatives and validated with examples from these initiatives. In addition, a Web-based platform is proposed as an implementation of the process [17]. It allows stakeholders to communicate, design, and share strategies for smart city development. The Web-based platform is based on the use of ontologies for a formal specification and representation of strategies' related knowledge. The proposed framework is then validated by a case study related to educational domain in a Tunisian city named Soussa [18, 19]. In the future work, we propose to add a simulation phase to the development process that we situate in between the phases of design and implementation. The simulation step evaluates the strategy in the city context before its implementation. It allows to test the proposed strategies on the city before implementing the strategy on the city. This will solve current problems of smart city strategies and gains time and money and allows to know if the strategy is suitable for the city.

References

1. United Nations., http://www.un.org/en/development/desa/population/publications/pdf/urbanization/the_worlds_cities_in_2016_data_booklet.pdf, (2016), accessed 14 June 2017
2. G. Maccani, B. Donnellan, M. Helfert, A comprehensive framework for smart cities, in: *Proceedings of the 2nd International Conference on Smart Grids and Green IT Systems* (2013), pp. 53–63, ISBN: 978-989-8565-55-6. doi: https://doi.org/10.5220/0004374400530063
3. S.B. Letaifa, How to strategize smart cities: revealing the SMART model. J. Bus. Res. **68**(7), 1414–1419 (2015)
4. L. Anthopoulos, N. Blanas, Evaluation methods for e-strategic transformation, in *Government e-Strategic Planning and Management*, ed. by L. Anthopoulos, C. G. Reddick, (Springer, New York, NY, 2014), pp. 3–23
5. M. Angelidou, Smart city policies: a spatial approach. Cities **41**, S3–S11 (2014)
6. L. Albrechts, Strategic planning as governance of long-lasting transformative practices, in *Human Smart Cities*, (Springer, Cham, 2016), pp. 3–20
7. A. Stratigea, C.A. Papadopoulou, M. Panagiotopoulou, Tools and technologies for planning the development of smart cities. J. Urban Technol. **22**(2), 43–62 (2015)
8. A. Ojo, E.Curry, T. Janowski, Designing next generation smart city initiatives-harnessing findings and lessons from a study of ten smart city programs, in: *22nd European Conference on Information Systems (ECIS 2014)*, 2014, vol. 2050, pp. 1–14
9. J.R. Harms, Critical success factors for a smart city strategy, in: *Proceedings of 25th Twente Student Conference on IT* (2016)
10. A. Taamallah, M. Khemaja, S. Faiz, Strategy ontology construction and learning: insights from smart city strategies. Int. J. Knowledge-Based Dev. **8**(3), 206–228 (2017)
11. Smart City Wien 2050., https://smartcity.wien.gv.at/site/files/2016/12/SC_LF_Kern_ENG_2016_WEB_Einzel.pdf (2015), accessed Nov 2017
12. New Castle Strategy., https://www.newcastle.nsw.gov.au/getmedia/392db4be-d418-48d8-a593-7a17a4b482bb/Newcastle-City-Council-Smart-City-Strategy-2017-21.aspx (2017), accessed 30 June 2018
13. London Strategic Plan 2015–2019., http://www.london.ca/city-hall/Civic-Administration/City-Management/Documents/COL_StratPlan_Brochure.pdf, accessed Nov 2017
14. Homeless Prevention and Housing Plan 2010–2024., http://www.london.ca/residents/Housing/Housing-Management/Documents/2017-09-05%20HomelessPreventionandHousingPlan.pdf, accessed Nov 2017
15. Planning Tools. London: Overseas Development Institute, https://www.odi.org/publications/5258-planning-tools-problem-tree-analysis (2009), accessed 15 June 2018
16. P. Gupta, M.C. Govil, Spring Web MVC Framework for rapid open source J2EE application development: a case study. Interface **2**(6), 1684–1689 (2010)
17. A. Taamallah, M. Khemaja, S. Faiz, A web-based platform for strategy design in smart cities. Int. J. Web Based Communities **15**(1), 62 (2019)
18. A. Taamallah, M. Khemaja, S. Faiz, Toward a framework for smart city strategies design, in: *Proceedings of the 3rd international conference on smart city applications* (p. 56). ACM (2018).
19. A. Taamallah, M. Khemaja, S. Faiz, Towards a framework for participatory strategy design in smart cities, in: *The proceedings of the 3rd international conference on smart city applications* (pp. 179–192). Springer, Cham (2018)

Chapter 4
Social Media as Tool of SMART City Marketing

The Role of Social Media Users Regarding the Management of City Identity

Dagmar Petrikova, Matej Jaššo, and Michal Hajduk

1   Introduction

Competition and cooperation process of the European cities and regions is considered as one of the most important scopes of the scientific discourse among spatial planners, economists, urban sociologist, and other experts. Fragile spatial and societal structures of the cities have been exposed to huge pressure originated either from international markets, unfavorable demographic prognosis, environmental hazards, or another source of risk (see, e.g., Finka et al. 2018 [1]). The cities and regions discovered completely new roles and positions within this process and they are learning how to take advantage from various factors. Globalization of the world accelerated this competition among the cities. European cities and regions are in the permanent process of aggravating competition in various markets. In order to be successful in this competition, the cities and regions have to thoroughly search for their competitive advantages (entirely analyzed in, e.g., Giffinger [2]). Recent development has shown that hard factors (geographical position, climate, density of population) are becoming less and less important and the cities are learning to search for the edge in the field of (so-called) soft factors: they are strategically managing their marketing activities, fostering social participation and involvement of the public, and promoting the leadership of their representatives and decision makers. Ability of cities to learn, adapt, and adjust to hyperdynamic outer and inner environment is a very important feature of any SMART project or initiative. It is based upon the repeated experience that successful SMART city or strategy cannot be completely emulated, copied, or artificially built but it has to take into consideration the wide range of specific traits, unique peculiarities, or distinctive characteristics of any given city.

D. Petrikova · M. Jaššo · M. Hajduk (✉)
Institute of Management, Slovak University of Technology, Vazovova, Bratislava, Slovakia
e-mail: dagmar_petrikova@stuba.sk; matej.jasso@stuba.sk; michal.hajduk@stuba.sk

© Springer Nature Switzerland AG 2020
N. V. M. Lopes (ed.), *Smart Governance for Cities: Perspectives and Experiences*,
EAI/Springer Innovations in Communication and Computing,
https://doi.org/10.1007/978-3-030-22070-9_4

City might be perceived as a metaphorical personality with various characteristics and traits. Place attachment of people is based upon the spontaneous mechanisms (memory, sense of togetherness, and pride) but might be actively fostered and supported by the various communication channels of the city. Construct "City as a Personality" might be a helpful tool in this field (see, e.g., Jaššo, Ladzianska 2015 [3]). SMART means unique and specific—in this context more than in any other.

SMART city is no longer perceived as mere highly advanced technological power, but the semiotic connotation of this label is nowadays leaning more toward the abovementioned soft factors. Unlike in the times of early modernity, current SMART cities need rather divergent than convergent creativity, "the creativity of being able to synthesize, to connect, to gauge impacts across different spheres of life, to see holistically, to understand how material changes affect our perceptions, to grasp the subtle ecologies of our systems of life and how to make them sustainable. We need skills of the broker" [4]. We are witnesses of the shift from fulfilling the complexity and hierarchy to ability to deliver uniqueness and specificity. Caragliu and Nijkamp [5] stated, "a city can be defined as 'smart' when investments in human and social capital and traditional (transport) and modern (ICT) communication infrastructure fuel sustainable development and a high quality of life, with a wise management of natural resources, through participatory action and engagement." Giffinger et al. [6] consider the main scope of smart city as a rather broad array of essential "characteristics," i.e., smart economy, people, governance, mobility, environment, and living. This broad view of smart cities has been anticipated by Manuel Castells even before the official concept of smart cities appeared (see Barth et al. 2017 [7]). Main feature of any smart solution in urban milieu is, according to Castells, the networking and connectivity. It is conclusive that almost any recent definition is leaning rather on some inter-relational, contextual, and/or soft variables (networking, communication, connectivity, human resources) than on strict delimitation of relevant technological factors/solutions.

In many cases, the city itself is being perceived and considered as unique organization, with various segments of target groups (visitors, investors, inhabitants), but also with its own stakeholders (citizens, companies, cultural bodies). The city strives to maintain its identity, to achieve ambitious goals, and to secure new developmental impulses and perspectives. Cities adopt the language and practice of commercial business practice [8, 9] and are obliged to create/discover and maintain their own corporate identity [10], figure out own place branding, and learn the new innovative ways of city marketing. Corporate identity of the smart city cannot be perceived as pure visual style or city design, but is rather a synchronized management of the "self" of the city. This concept (CI) should cover and organize in a formidable way all the elements of the city identity: visual style, communication, and behavioral patterns. The most efficient way to measure the efficiency of the corporate identity of any given city is to evaluate the authenticity, trustworthiness, and stability of its image. Image of the city is thus overlapping its visual nature, being rather mental construct than pure visual imaginary.

Despite many differences when comparing the territorial subjects (city, region) and commercial subjects (company), one feature is strikingly similar: need for active, humane, and transparent communication with external environment. City is a living organism and its communication is neither propaganda nor purely political message; it is a dialogue/polylogue led by many different actors and stakeholders. The process of this communication is neither purely accidental nor purely determined: city is evolving during the communication process and is thus a kind of "self-learning system."

Communication of the city might be defined as a continual and long-term process of information exchange aimed to satisfy needs of all participating parties. Any external cooperation of the city should be based upon clear, transparent, and effective communication process and requires highly profiled communication skills of the actors of communication. These requirements are concerning the communication techniques and methods, effective utilization of communication channels, as well as presence of ethical standards and principles. Ethical communication is not held in strictly outlined frame sender-receiver, but it is producing a very dynamic system of mutual feedbacks between various communication partners. Generally, it is not communicated something to someone, but with someone about something. This might be fully in concordance with the definition of "smart communication."

Transmitting this into municipal practice, this outcoming from darkness of anonymity means accepting the responsibility and preparing the conditions for a dialogue or polylogue—as a qualitative higher form of communication. This way of communication is an open system—it does not have ambitions to find so-called perfect solutions, and does not force the terminal results as a final phase of communication. The understanding and meaningful relation toward the public are evaluated as the wished result and this moves the whole process to superior levels. The feedback and the "revival of words and feelings" are the basic tools of this kind of communication.

There are many specific characteristics of stakeholders/target groups in the process of external communication of the city. The following ones we consider to be most essential:

(a) Close connection to target audience, mainly toward the inhabitants: Issues related to the city are concerning almost every citizen and are communicated in high frequency. Many topics bear high emotional burden and might be highly sensitive for many actors.
(b) Immense variability of target groups (enterprise bodies, citizen initiatives, gastronomy, tourism): It is necessary to harmonize the interests of the rather external (investors, tourists, media) and rather internal (citizens, local entrepreneurs, youngsters) audience. The goals leading to satisfaction of the needs of both groups may not be in contradiction. That is the reason why the detailed and deep analysis of the market as well as the analysis of the needs and expectations of the various target groups should be in the focal point.
(c) Mutual information exchange and willingness to communicate among the participants of the process: This is necessary to be taken into account if the coordination of the realized measures and maximalization of positive effect have to be guaranteed. If the particular segments of target audience are not communicating together, the effective communication of the city is almost impossible.

(d) We have to bear in mind that many steps in city marketing and external communication are not reversible (unlike in the commercial business marketing) and might influence the future identity of the city for many decades. City must be very cautious in many particular aspects and steps of the entire communication process.

Each city should develop its value profile, certain typical set of principles and beliefs, transformed into the behavioral preferences and communication patterns. One of the most appreciated benefits is loyalty toward the city, and stable, long-term, and deeply profiled set of the attitudes of the public and other city partners. Very important part of this concept is sense of belonging and place attachment of the inhabitants and other actors in city life. Highly profiled smart city identity and strong ties of place attachment are of utter importance for social cohesion within the territory and territorial aspect is a crucial dimension in the concept of social identity and sense of belonging. Place attachment saturates many psychological needs: the need for security, the need for self-realization, the need for belonging, and the need for structuring the outer environment (see, e.g., Jaššo, Petríková 2015) [11].

In many successful smart city projects and strategies, there is encapsulated an underlying assumption that city is a learning system. The increased importance of the ethic criteria of the communication process in the city management is outgoing from the previous statements. Only if the given city is smart/intelligent the consequences from previous behavior can be taken into consideration and thus the system/city itself is able to correct, modify, and evaluate its inner potential. The behavioral autonomy, high ethical standards, and strongly profiled identity are inseparable parts of these systems. The maintenance and development of these fields is the principal platform for the development of unique and memorable city identity.

2 New Trends in Commercial Business Regarding the Social Media

As the cities borrow the knowledge from the commercial business practice it is important to investigate the new trends in terms of marketing, branding, and corporate identity triggered by the widespread use of the Internet and most important by the phenomenon of Web 2.0. Web 2.0 can be defined as a "second generation of Internet services that are based on the online collaboration, the interactivity and the ability to share content among users. Technological development has allowed a much deeper change than programming languages and tools of content publishing. For this reason we speak of the 2.0 phenomenon as something that transcends the barriers of technology, it is a new paradigm of relationship and knowledge, similar to that came with the advent of printing" [12]. The fast development of Internet possibilities led to the emergence of the so-called social media sites and networks which group masses of users together and give them opportunity to communicate constantly in real time regardless of the place. Such services do offer not only communication, but also possibility to create own content and share it with the world. Users are able to interact with each other and form communities based on their interest where they share experiences and opinions. These platforms enhance and multiply social experiences of their users and can "fulfill their social value by satisfying their need for belongingness and their need for cognition with those who have shared norms, values and interests" (Gangadharbhatla, 2008) [13]. Social media created online social sphere which coexists and blends with the real world. Regarding the commercial business "the so-called social media let users to have an active role, creating directly informatics contents shared in an easier way than the past, sometimes being able to catch the attention of the public opinion, reality of corporate life that before was hidden or difficultly accessible and undisclosed, as well as collect sentiments and opinion about product and corporate experiences" [14]. Commercial business recognized the advantages of closer relationship with customers and learned how to listen to their needs and desires. Such data provide valuable information about the market and possibility to adjust the product or service to the needs of their clients. Social media and the phenomenon of Web 2.0 "have increased interest in the scientific community and practice management for the new opportunities and challenges to manage the corporate reputation using the web itself" [20]. As Labajos and Jimenez-Zarco point out "although you cannot control what happens in social media, you can learn from them and design the most appropriate message to reach the audience effectively" [15]. There is an evident shift in the commercial business from traditional product/service marketing to a social marketing. Such "social commerce gives the company the opportunity to know its customers, adjust its offer and generate useful experiences that satisfy the demands of consumers, fostering a relationship with the brand's users" [15]. According to Labajos and Jimenez-Zarco: "regarding the 2.0 phenomenon is the new mentality adopted by companies, putting themselves on the same plane as their stakeholders and allowing these stakeholders to have a role in the business processes of the company" [15]. Castelló and Ros summarize that the relevance of the Internet has increased to such an extent that online actions are now a fundamental pillar in any brand's communication strategy [16].

From abovementioned statements, it is evident that the social media became a game-changing stimulus to a way how is the customer approached by the commercial business marketing and how is the customer becoming an essential part of such processes.

3   Social Media and the City Marketing

The presence of dwellers/inhabitants/residents of certain areas on social networks presents possibility to directly address these communities and potential to strengthen their place attachment as well as their sense of togetherness and social cohesion. According to Balmer "The feelings of social and cultural connection tie people to a community, and such community can be enhanced through social media" [17].

Social media offers possibilities to bring together people with similar interests and opinions or with shared living space, and is laso able to foster the social identity of each user, which "implies positive feelings toward being a member of the group and emotional involvement in the group's activities" [18]. Social identity, as Ukpabi's and Karjaluoto's work on social media suggests, "also motivates participation toward the achievements of the interest of the group. Therefore, social media as a platform fosters that collective interest by providing a platform for members to share and advance the brand" [18].

Another aspect of social media is that "it has the characteristic of information openness, participation, interaction, sharing, connectedness, creativity, autonomy, collaboration and reciprocity" [19] and therefore "online social networks represent progress for democracy, because they allow everyone to take part in public debate" [20]. Unlike traditional media "where journalists or editors can decide what information will be delivered to the audience" [21], comments and feedback are provided directly by the citizens. The potential of social media could bring more transparency to the city administration and its processes. As the "most valuable feature is the possibility of direct, two-way communication between the content creator and the user" [21], social media could help municipalities communicate their activities and strategies and involve citizens in participation. "Involved in this process, they influence the directions of the city renewal and support the city authorities and urban planners with suggestions and conclusions coming from experiences gained in real urban space and not from studies of the maps and reports" [22]. As Lusoli, Ward, and Gipson point out "it will require a demonstration that their participation and communication is valued and listened to" [23] as well as "the dialogue needs to be ongoing, considerably less top-down and less formalized" [23]. Compared to traditional communication of the cities' administration, social media provides dialogue, in which "the consumer is no longer the passive recipient of marketing messages but rather is very active in controlling and co-building corporate brand" [24]. According to many studies "using social media for political purposes does have a positive influence on political interest and offline political participation" [25], and also "social media can be used to enhance youth political participation" [26] and "social media have the potential and ability to promote online civic participation" [27].

Regarding the city marketing the social media provides invaluable opportunity to engage the citizens to participate in the effort and help achieve its goals. According to Labajos and Jimenez-Zarco [15] social media transformed the former customer (citizen) to a new type of recipient of the marketing message. Authors divide the customers into three different roles they play on the Web: prosumer, crossumer, and adsumer.

The prosumer, composed of terms PROducer and conSUMER, "is the new consumer, the one who interacts with the Web and creates trends, influencing other users. What really makes the difference on the Web 2.0 is that users now have an active role, as they share information, talk, analyze, discuss, claim, innovate and research" [15]; they act as human communication channels. In the social sphere of social media they operate as the influencers.

The crossumer is defined "as an expert consumer that decodes the intentions of advertising campaigns and branding strategies, rejecting the communicative mono-logue of the great campaigns" [15]. According to Neumeier [28] the crossumer is familiar with the marketing and its message and decodes the communication and questions the meaning. Such a consumer is experienced in the search for informa-tion and their verification. He/she crosses information sources in order to validate or discard the message. His/her opinion contributes to the other user-generated feed-back on the message which increases or reduces the credibility of the information. Such an actor is committed; if he/she likes the message he/she joins in and if he/she does not he/she is ready to boycott the message.

The adsumer, composed of terms ADvertising and conSUMER, can be charac-terized as "the satisfied customer who recommends a brand on the Net, basing his opinions on his own previous experience" [15]. Based on their negative experience they may loudly complain about the product, service, or company. According to Castelló [29] the extreme version of adsumer is the FANsumer which is "the indi-vidual who identifies himself with the values of the brand to the point of having a great empathy for it. They are individuals who evangelize about the brand, defend it and propagate virally any content related to it."

The shift from one-way communication of the previous generation of the Web to the two-way communication pushed the Internet users from the role of percipient to the role of active co-creator of the online content. Therefore they are no longer just percipients regarding the place branding and corporate identity of the city, but become the active agents that follow the happening in the city and are ready to pro-vide critique to any inconsistencies in the communicated message.

Abovementioned literature review suggests that the social media widespread thanks to the phenomenon of Web 2.0 could be used as a useful tool for the munici-palities regarding the aims of the city marketing. Therefore we assume that social media has the potential to foster communities, help to formulate and communicate shared believes and aims, and involve people to participate, if the message is truth-ful and worth acquiring. "If the company learns to listen to and influence the public it can assure itself a market position with very strong social acceptance (to influence the public means to provide experiences that will incite them to speak positively of the company, not to deceive, control or coerce)" [15].

4   The Use of Social Media by Municipality of Bratislava and the Citizen's Interest Groups/Urban Activists

Bratislava, the capital of Slovak Republic, was selected as a case study for the usage of social media by the municipality officials and the engaging public in terms of city development. Bratislava is a relatively small capital with approximately 421,000 inhabitants and similarly to other cities in the Central Europe region struggles with defining its own identity in the present.

Bratislava has launched its current corporate identity in 2004. The project has been largely reduced to pure visual style, based upon the simplified symbol of Castle of Bratislava and the claim "Little Big City." After initial critics, the visual style has been accepted relatively smooth by public and other target groups, but the overall concept of corporate identity still lacks other substantial elements: namely the city communication strategy and guidelines for city behavior. Bratislava also participated in other regional marketing projects and activities; the concepts of Twin City (with Vienna) and Region CENTROPE are worth mentioning.

For better understanding of the current state of the usage of social media regarding the city development and management, we managed to collect information from the Facebook pages administered by the municipality and by the mayor of Bratislava. These two Facebook pages are official pages for the municipality of Bratislava—*"Bratislava—hlavné mesto SR"* (Bratislava—the capital of SR), and the official page for current mayor of Bratislava—*"Ivo Nesrovnal pre Bratislavu"* (Ivo Nesrovnal for Bratislava).

All mentioned Facebook pages are active and reached several thousand followers that actively contribute to the discussion about the published posts (to the date of 24.06.2018):

Facebook page name	Number of followers
"Bratislava—hlavné mesto SR" (Bratislava—the capital of SR)	18,599
"Ivo Nesrovnal pre Bratislavu" (Ivo Nesrovnal for Bratislava)	15,232

Studied Facebook pages were selected as examples of official political stakeholders in the city using social media. As there are many online pages used by the municipality, dedicated to each district of the city or the transportation, etc., and many pages used by the urban activists and citizens, selected pages could provide reliable information about the usage of the social media regarding the city development and management by official city representatives. The comparison of the data obtained from mentioned Facebook pages could provide valuable information about the current state of the digital realm of the city and the perception of such activity by the citizens and its impact on the physical realm of the city.

5   Methodology

Our study of social media as a tool for SMART city marketing and management consists of two main parts. First part of the study focused on quantitative/qualitative gathering of the data provided on the social network Facebook by the two selected Facebook pages in the time span of 6 months. In the case study of "Bratislava—hlavné mesto SR" page in the period from 1.1.2017 to 1.7.2017, for the other case study of "Ivo Nesrovnal pre Bratislavu" (Ivo Nesrovnal for Bratislava) page in the period from 1.1.2018 to 1.7.2018. The study focuses on the public posts published by selected entities and the data obtained from such posts are categorized by the method of communication, topic of the published message, engagement of the followers, and the type of feedback provided by the followers/public.

Method of providing the message was categorized as:

1. Text
2. Text + graphics, photographs, video, or other audiovisual material
3. Text + link to the external source of content

Topic of the messages was divided into 7 main categories and 23 subcategories that prevailed in the posts:

1. City infrastructure

    (a) Public spaces
    (b) Greenery, nature
    (c) Traffic, public transportation, cyclist
    (d) Visual smog, graffiti
    (e) Safety
    (f) Waste management

2. Announcements

    (a) Warnings, limitations
    (b) Danger

3. Events, happenings, tourism

    (a) History
    (b) Sport
    (c) Culture
    (d) Other events

4. Activities of municipality

    (a) Small urban improvements
    (b) Large urban improvements
    (c) Funding, budget, finances
    (d) Planning, strategies
    (e) Activities of city council
    (f) Housing

5. Participation

    (a) Participation with citizens

6. Volunteerism

    (a) Organized by the municipality
    (b) Organized by activists and other groups

7. Entertainment

   (a) Competitions
   (b) Entertainment

Engagement of the citizens was expressed by the number of:

1. Likes (and similar emoticons)
2. Shares of the posts
3. Comments

Content of the comments was divided into three categories as:

1. Questions to the municipality/mayor
2. Questions answered by municipality/mayor
3. Comments and opinions about the post

The attitude expressed in the comment section by the public was categorized as:

1. Positive
2. Negative
3. Neutral

Second part of the study consists of comparative analyses of gathered and categorized data from two different Facebook pages. Comparison between the pages will focus on the use of social media in terms of digital city management and marketing by different subjects, frequency of the publishing, topics of published posts, and effectiveness of using social media in terms of engagement of the public and attitude of percipients.

6 Case Studies

6.1 Page: "Bratislava—hlavné mesto SR" (Bratislava—the Capital of SR)

Analyzed data shows that most of the Facebook posts supported the message by the graphical material in the form of photographs or videos (53.6%), but also that considerable amount of posts simply referred to the content linked to other source (44.2%) and only insignificant amount of messages were provided just by plain text (2.2%) (Fig. 4.2).

The topic of the posted content referred mostly to the city infrastructure (32.4%); events, happenings, and tourism (22.5%); and the activities of the municipality (21.6%). Announcements (9%) and entertainment (8.4%) were present with similar amount of posts. Least number of posted messages were about volunteerism (3.5%) and really insignificant space was represented by the topic of participation with citizens (2.6%) (Fig. 4.1).

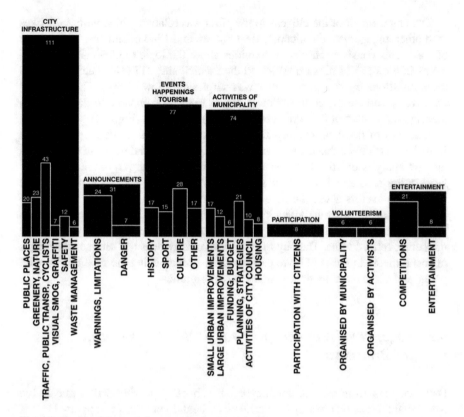

Fig. 4.1   Topics of the posted content by the municipality

Fig. 4.2   Type of provided message and the engagement of the citizens

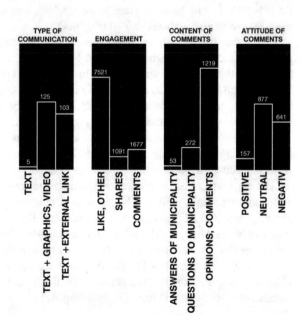

The engagement of the citizens to the posts was relatively high with 7521 likes (and other engagement emoticons), 1091 shares, and 1544 comments. The content of comments consisted mostly of opinions about the topic (78.9%) and the comments in the form of direct question to the municipality (17.6%). The response to these questions by the municipality was surprisingly low (3.5%). Therefore, the negative attitude in the comments was higher (38.2%) compared to positive comments (9.4%). Most of the comments were neutral (53.4%) (Fig. 4.2).

Analyses of the data obtained from the Facebook page of the municipality of Bratislava has shown that the presence of the city on the social media in this particular case study is practically targeted on pure informational purpose and is clearly lacking the substantial definition signs of SMART city. Facebook page mostly informs the visitors about the state of urban infrastructure or reminds them of the cultural events or other activities in the city. There is almost no information about the strategies and planning including targeted mention of the city identity or any city marketing-related topic. The city of Bratislava clearly does not use the advantages of social media in terms of civic participation, strengthening the relationship with public and place attachment or marketing opportunities.

6.2   Page: "Ivo Nesrovnal pre Bratislavu" (Ivo Nesrovnal for Bratislava)

Data obtained from the current mayor's Facebook page shows that mayor Ivo Nesrovnal focuses on creating own original content, in most cases supported by the graphical materials in the form of pictures and video (85.8%), while only a few posts were referring to an external source (8.5%), and also the messages shared in plain text were also avoided (5.7%) (Fig. 4.4).

City infrastructure was the most shared and discussed topic during the studied period of time (36.2%), which is linked to the mayor's large road repair program. Second most posted topic was referring to the activities of the municipality (28.8%), mostly to the strategies and planning of the city. City events, happenings, and tourism was the third most posted topic (23.7%) with the focus on culture and sport events. There were few announcements regarding the limitations or warnings for the citizens (4.7%), and entertaining post or competitions (3.1%), but only very small amount of posts regarding the volunteerism (2.3%) and insignificant number of posts regarding the participation processes in the city (1.2%) (Fig. 4.3).

The engagement of the citizens to the messages posted by the current mayor of Bratislava was high with 15,760 likes (and other engagement emoticons), 1027 shared posts, and 3276 comments. The content of comments consisted mostly of opinions about the topic (81.1%) and the comments in the form of direct question to the mayor (8.6%). The response to these questions by the mayor was nearly as high as the number of questions (7.3%). Negative attitude in the comments was higher (24.9%) compared to positive attitudes in the comments (17.5%). Most of the comments were neutral in tone (57.6%) (Fig. 4.4).

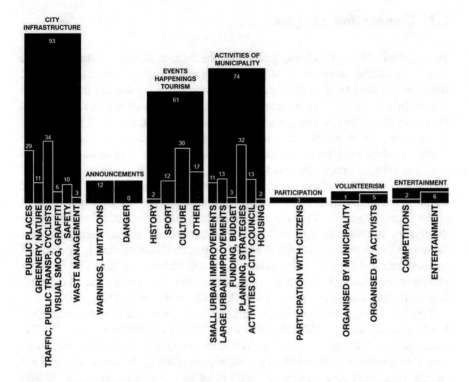

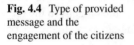

Fig. 4.3 Topics of the posted content by the mayor

Fig. 4.4 Type of provided
message and the
engagement of the citizens

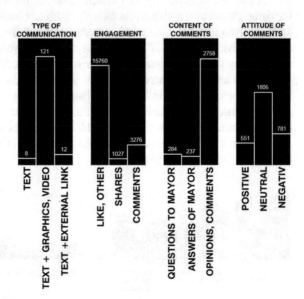

6.3   Comparative Analyses

Two studied official Facebook profiles of the current Mayor of Bratislava Ivo Nesrovnal and the municipality of Bratislava in many cases provided similar data. Both profiles were focused mostly on information regarding the city infrastructure (36.2% mayor; 32.4% city) and the most shared and discussed topics for both profiles were about the transportation and the cyclists. Second and third most shared topics were for both profiles also similar—activities of the municipality (28.8% m; 21.6% c), where both profiles mostly focused on the topic of planning and strategies, and events, happenings, and tourism (23.7% m; 22.5% c) where the cultural events were the most discussed. Announcements (4.7% m; 9% c) in the form of warnings and information about danger or limitations in the city were more present on the Facebook profile of the municipality, which was more focused on informing the citizens than the mayor. Municipality also shared more entertainment posts than the mayor (3.1% m; 8.4% c), which tried to engage with the citizens in the form of competitions and humor. Volunteerism (2.3% m; 3.5% c) was also slightly more present in the Facebook feed of the municipality, but both profiles shared information about the participatory processes the least (1.2% m; 2.6% c) (Fig. 4.5).

Differences begin to appear in the comparison of engagement, where the mayor Ivo Nesrovnal (with 15,232 followers) with 3367 less followers compared to the municipality of Bratislava (with 18,599 followers) gained nearly two times more overall engagement by the public than the municipality. In the terms of likes (and similar emoticons) mayor engaged public twice as much as municipality (15,760 likes mayor; 7521 likes city), but posts published by mayor and municipality had similar number of shares by the citizens (1027 shares mayor; 1091 shares city). In the comment section both profiles gained significant number of comments (3279 m; 1544 c), from which similar number were direct questions of the citizens to the city representatives (284 m; 272 c), but the questions were much more likely to be answered by the mayor, than the municipality (234 m; 53 c) (Fig. 4.6).

Attitude toward the shared messages was mostly neutral in both cases (57.6% mayor; 52.4% city), but interesting is the negative perception of the shared posts where the information shared by the municipality was perceived much worse than the information shared by the mayor (24.9% m; 38.2% c). Nevertheless, the negative attitude in the comment section was higher than the positive attitude in both cases (17.5% m; 9.4% c) (Fig. 4.6).

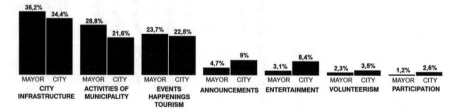

Fig. 4.5 Comparison of the topic shared in the posts by the mayor and the municipality

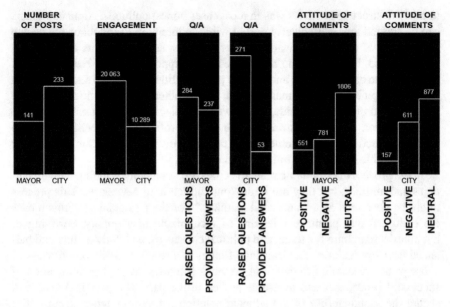

Fig. 4.6 Comparison of the engagement in the posts by the mayor and the municipality

7   Conclusion

Analyses of the data obtained from the Facebook page of the municipality of Bratislava and the Facebook page of current mayor of Bratislava have shown that the presence of the city and the city representatives on the social media in this particular case study is practically targeted on informational and self-promotional purposes and is clearly lacking the substantial definition signs of SMART city. Facebook pages mostly inform the visitors about the state of urban infrastructure or remind them of the cultural events or other activities in the city. In the case of current mayor, Facebook also serves as a platform for self-promotion, which is kind of expected considering the political function of the mayor. There is almost no information about the strategies and planning including targeted mention of the city identity or any city marketing-related topic. The city of Bratislava clearly does not use the advantages of social media in terms of civic participation, strengthening the relationship with public, and place attachment or marketing opportunities. Better use of the Facebook is on the side of current mayor Ivo Nesrovnal who clearly understands the current state of social media better and tries to take advantage of them to gain more political points. Bigger engagement of the public on the Facebook profile of the mayor could be explained simply by the fact that people are more likely to engage with a concrete public figure, than with the faceless profile of the whole municipality. Therefore people are more likely to obtain answers to their questions and even build more personal relationship with such a figure. Even though the citizens' engagement from really positive to fairly destructive attitudes was in most cases honest opinions, some comments can be spotted that were clearly posted with the intention to influence the debate in the favor of the profile owner. Profiles that post such comments are probably paid to do so, but there is no real evidence to support this fact. Nevertheless it points out the dangers of the anonymous public online debate, and shows how can the public opinion be easily manipulated. On the other side the same effect can be brought from the opposing side, where well-organized groups can manipulate the public opinion against the owner of the online profile. Such a behavior can be really destructive and, in some cases, can destroy the whole conversation, or worse it is able to shape the opinion of the public.

In the process of communication, the cities are not merely utilizing their advantages and abilities, but they are improving and learning new skills. This process might be described as a "municipal learning." For the particular city, this means the continual and permanent lessons of recognizing new quality, verifying of hypothesis, searching for social and political consensus, and daily mature and balanced decision-making. The learning of cities is a result of deliberate process of consequence research, evaluation of previous activities, as well as application of successful procedures and measures proved in the past. The principal task is to secure the sustainability of the adopted solution for various target groups. The immediate and full feedback as well as the permanent monitoring of running activities are of high importance, particularly with regard to spatial planning activities. Many measures and steps do not enable a withdrawal from planning practice, due to their irreversible nature. It is necessary to anticipate the future municipal development, to judge and estimate the risk potentials and the possible escalation focal points, to reveal hidden sources, and to utilize them in a friendly, open, and honest manner. As Lovari and Parisi based on their study point out, "public institutions are managing their Facebook Pages in a way that is far from fulfilling the expectations of their fans" [30], and Gausis summarizes, "Institutions should pay more attention to their activities on social media if they want to get successful communication outcomes and not just being present on social media to demonstrate that they theoretically care about citizens and their opinions" [21]. What is worth mentioning is the movement of the citizens present on the social media which actively engage in the discussion and provide useful feedback for the city. Such public is naturally sensitive about the messages spread by the city, which is their living space. Any contradiction between the reality and the messages communicated by the city is quickly exposed and generates adequate response. Such a behavior observed in the case study fully justifies the assumptions based on the literature review. Regarding the possibilities given to people through the social media, a paradigm shift in the relationship between the percipients and broadcaster of the message is evident. The presence of active groups of citizens on the social media and the possibilities the social media nowadays present are invaluable tools for the municipalities.

7.1 Future Research

Aim of this chapter was to present the paradigm shift in the relationship between different actors, in our case the city and its inhabitants, that was fostered by the emergence of the social media and the possibilities that have been opened up. For the future research regarding the city management as a tool of spatial planning, there is a need to figure out the ways to take advantage of the social media in terms of creating or strengthening the urban identity and building the corporate identity of the city and to engage the public present on such platforms to cooperate to and participate in such actions. Another important topic would be how to use the social media for further marketing of the city and strengthening of the relationship with public, engaging public in further building and promotion of shared goals and aims, and how to adjust the processes according to the constant feedback and communication.

References

1. M. Finka, M. Husár, M. Jaššo (eds.), *The Role of Public Sector in Innovative Local Economic and Territorial Development (In Central, Eastern and South Eastern Europe)* (EAI/Springer Innovations in Communication and Computing, Cham, Switzerland, 2018)
2. R. Giffinger (ed.), *Competition Between Cities in Central Europe: Opportunities and Risks of Cooperation.* ROAD Bratislava (2005)
3. M. Jaššo, Z. Ladzianska, City as a personality: new concept of creative city in: *Smart City 360: 2nd EIA International Summit,* 22–4 November 2016, Bratislava. 1. ed. Bratislava: EAI online
4. C. Landry, F. Bianchini, *The Creative City* (Demos Publishing, London, 1995)
5. A. Caragliu, C. Del Bo, P. Nijkamp, Smart cities in Europe. J. Urban Technol. **18**, 65–82 (2011)
6. R. Giffinger, C. Fertner, H. Kramar, R. Kalasek, N. Pichler-Milanovic, E. Meijers, *Smart Cities: Ranking of European Medium-Sized Cities* (Centre of Regional Science, Vienna, Austria, 2007)
7. J. Barth, K.J. Fietkiewicz, J. Gremm, S. Hartmann, A. Ilhan, A. Mainka, C. Meschede, W.G. Stock, Informational urbanism. A conceptual framework of smart cities. in: *Proceedings from 50th Hawaii International Conference on System Sciences,* 2017
8. M. Konken, Realität einer Vision. Stadtmarketing—der exakte Weg ist das Ziel, in: www.kommunale-info.de/themen/stadtmarketing/organisation.htm, accessed 31 Oct 2001
9. P. Kotler, D. Haider, I. Rein: Standortmarketing. Wie Städte, Regionen und Länder gezielt Investitionen, Industrien und Tourismus anziehen. Duesseldorf, Wien (2004)
10. M. Jaššo, M. Hajduk, Strategický manažment identity mesta: corporate identity alebo place branding?/Strategic management of city identity: corporate identity or place branding? In: Podniková ekonomika a manažment, Nr. 2, pp. 22–32 (2016)
11. M. Jaššo, D. Petríková, Place attachment and social communities in the concept of smart cities, in: *Smart City 360°: First EAI International Summit, Smart City 360°,* Bratislava, Slovakia and Toronto, Canada, October 13–16 2015, Bratislava. Revised Selected Papers. Vol. 166. International conference on Sustainable Solutions Beyond Mobility of Goods, Sustainable MoG 2015, Bratislava, Slovakia, 13–14 October 2015 (2016).
12. R. Levin, C. Locke, D. Searls, D. Weinberger. The Cluetrain Manifesto. In: http://www.cluetrain.com/book/ (1999)

13. H. Gangadharbhatla, Facebook Me: Collective self-esteem, need to belong and internet self-efficacy as predictors of the I-generations attitudes toward social networking sites. J. Interact. Advert. **8**, 5–15 (2008)
14. S. Alfiero, C. Massimo, P. De Bernandi, V. Tradori, Social media corporate reputation index. How social influencers affect on corporate reputation, Impresa Progetto Electronic J. Manag. (2) 2016
15. N. Soler-Labajos, A.I. Jimenez-Zarco, The valuable alliance between social media and e-commerce: social networks as a tool for transparency, dialogue, and sales (2016)
16. A. Castelló, V. Ros, CRS communication through online social media, in: *Revista Latina de Comunicacion Social* (2012)
17. J.M.T. Balmer, M. Burghausen, Explicating corporate heritage, corporate heritage brands and organization heritage. J. Brand Manag. **22**(5), 364–384 (2015)
18. D. Ukpabi, H. Karjaluoto, Influence of social media on corporate heritage tourism brand, in: R. Schegg, B. Stangl (eds.), *Information and Communication Technologies in Tourism* (2017)
19. J. Nancy European Youth in 2016. European Parliament. Retrieved from http://www.europarl. europa.eu/pdf/eurobarometre/2016/eye2016/eb85_1_eye_2016_analytical_overview_en.pdf (2016).
20. Y. Jiao, M.S. Jo, E. Sarigollu, Social value and content value in social media: Two paths to psychological well-being. Journal of organizational computing and electronic commerce **27**(1), 3–24 (2017)
21. E. Gausis, European institutions on social media: shaping the notion of European citizenship, in: De Gruyter 30(1) published online 12.4.2017 (2017)
22. N. Urbanowicz, Media architecture and interactive art installations stimulating human involvement and activities in public spaces, in: *CBU International Conference on Innovations in Science and Education*, Praha, 23–25 Mar 2016, www.journals.cz (2016)
23. W. Lusoli, S. Ward, R. Gibson, (Re)connecting politics? Parliament, the public and the internet. Parliamentary Affairs **59**(1), 24–42 (2006). https://doi.org/10.1093/pa/gsj010
24. A. Rindell, T. Strandvik, Corporate brand evolution: corporate brand images evolving in consumers' everyday life. Eur. Bus. Rev. **22**(3), 276–286 (2010)
25. K. Holt, A. Shehata, J. Stromback, E. Ljungberg, Age and the effects of news media attention and social media use on political interest and participation: do social media function as leveller? Eur. J. Commun. **28**(1), 19–34 (2013). https://doi.org/10.1177/0267323112465369
26. L. Vesnic-Alujevic, Young people, social media and engagement. European View **12**(2), 255–261 (2013). https://doi.org/10.1007/s12290-013-0282-2
27. M.A. Warren, A. Sulaiman, I.N. Jaafar, Social media effects on fostering online civic engagement and building citizen trust and trust in institutions. Gov. Inf. Q. **31**, 291–301 (2014). https://doi.org/10.1016/j.giq.2013.11.007
28. M. Neumeier, *The Brand Gap: How to Bridge the Distance Between Business Strategy and Design* (New Reader, Berkeley, 2014)
29. A. Castelló, Crossumer, prosumer, fansumer y persumer. Estrategias empresariales en social media. Retrieved from http://www.aracelicastello.com/2010/07/crossumerprosumer-fansumer-y-persumer.html (2010)
30. A. Lovari, L. Parisi, Listening to digital publics. Investigating citizens' voices and engagement within Italian municipalities' Facebook Pages. Public Relat. Rev. **41**, 205–213 (2015). https://doi.org/10.1016/j.pubrev.2014.11.013

Chapter 5
Inclusive and Accessible SMART City for All

Dagmar Petríková and Lucia Petríková

1 Introduction

Urbanisation and growth of cities is a serious and long-standing trend. The social and economic inequalities are most visible in cities. According to WHO, the urban population is globally predicted to climb to 70% by 2050 [1]. The world population is predicted to reach 9.3 billion people by 2050 and at the same time most countries will reach a highest increase in ageing population. By 2030, 25% of the European population will be over 60 years old; meanwhile the overall population is predicted to drop [2]. With the ageing population comes a higher percentage of health-related issues, impairments as well as disabilities. As a result of the increasing number of urban population, it is urgent to build cities that keep in mind special needs of their citizens, particularly the most vulnerable of them—the disabled and the elderly.

Urban settlements and especially urban centres are highly differentiated and involve various social groups of citizens with their specific lifestyle needs, interests, attitudes and ability to move around in space. The city which directs its development towards a higher quality of life and moves towards a smart city must reflect the needs of specific population groups that react much more sensitively to situation and conditions in areas, especially public spaces. This chapter examines the smart city concept from the perspective of specific groups of citizens living there. It provides the overview of different interpretations of smart cities and identifies the target groups and their requirements for the quality of space.

D. Petríková (✉) · L. Petríková
Institute of Management, Slovak University of Technology in Bratislava,
Bratislava, Slovak Republic
e-mail: dagmar_petrikova@stuba.sk; lucia.petrikova@stuba.sk

© Springer Nature Switzerland AG 2020   73
N. V. M. Lopes (ed.), *Smart Governance for Cities: Perspectives and Experiences*,
EAI/Springer Innovations in Communication and Computing,
https://doi.org/10.1007/978-3-030-22070-9_5

2 Sustainable Development in SMART City for All Concept

To support inclusive, sustainable urban development in smart city concepts, it is necessary to demand high-quality public space accessible for all citizens. By public space we understand spaces in urban infrastructure that are open and accessible for all citizens regardless of their physical and mental abilities, age, gender, race, ethnicity or socio-economic level. A quality public space is able to meet a variety of different users' needs. People with special needs are not characterised by a homogenous group of users but they consist of individuals and groups with different requirements for the quality of urban environment, especially public spaces. Starting from a position of people's capabilities rather than their disabilities, the previous research has outlined the following principles addressing a wide range of quality attributes of public space [3–5].

2.1 Social Interaction

The urban environment must reflect the diversity of people who use it and provide a place for social interaction. The more inclusive the space is the more social interactions occur. A quality urban design offers space that people use to form strong, vibrant and sustainable communities. A quality space is designed with everyone in mind. Therefore, it is essential to involve as many people as possible during planning processes, not excluding the disabled people, seniors and families with small children. Well-maintained public space and green areas are good for health, wellbeing and community life. Inclusive public space is open, safe, accessible, functional and enjoyable to use. It is a place that enhances interaction between different social groups, where people of different age can meet and where various activities occur. It is designed with wide and smooth pedestrian paths for easy movement of different mobility impairments. It provides a shady, secure place for older people; has a good lighting and clear orientation signs; is designed with universal equipment; and is clean and well maintained.

2.2 Urban Safety

One of the most important attributes of the quality of space is the urban safety. The urban safety relates to any kind of user safety in relation to the urban space, especially in public areas. It includes a wide range of issues related to the safety of the physical environment such as barrier-free accessibility, crime prevention and universal design principles [3, 4]. A quality public space should be easy to get to and through. If people have difficulties to get to and around a place they will probably not (want to) use it at all. Inclusive urban planning must recognise barriers that people are mostly affected by. People with special needs need places which fit a purpose and are easy to understand and convenient to use. People suffering from mental illnesses or sensual impairments often struggle with a design that is confusing and counter-intuitive. A safe urban public space works with signage, lighting, visual contrast and materials; has simple and clear street signs (directional indicators) to help people with sight, hearing or mobility impairments to orientate; and has sidewalks with clear sight lines to make women, seniors and teenagers feel safe. It is designed with citizens' participation. It creates neighbourhoods that are designed with easy orientation to help people with disabilities find a way around. Technical attributes such as ramps and toilets are important but always in relation to the accessibility, safety and quality of the whole space. Urban safety also means well-connected streets that connect public spaces so that vulnerable people feel safe.

2.3 Accessibility

To meet accessibility requirements of all users it is essential to implement standards that treat everyone on equal terms. The accessibility of any place is defined by its visual and physical connections to its surrounding. Accessible public spaces are easy to get by means of public transportation, cycling or walking. Accessibility of public space depends also on parking options for disabled, walkability of a place, distance from public building entrances and barrier-free access options. For people with special needs it is necessary to have sufficient information (possibly even before leaving their home) about public transportation timetable, barrier-free possibilities, opening hours and public services. An accessible public space is not occupied by parking cars, has pedestrian paths that connect adjacent areas, provides people with a variety of public transportation options and has stops comfortably located in proximity to a place.

2.4 Mobility

While it is necessary to address the needs of wheelchair users and other mobility-impaired people, smart cities also need to recognise problems of people with learning difficulties, mental illnesses, and visual and hearing impairments when it comes to inclusive and accessible mobility and movement in a space. A flexible mobility consists of a well-designed, barrier-free public transportation system and walkable zones in urban areas which are intuitive and easy to orientate in. Inclusive public transportation system provides options for everyone regardless of their age or physical or mental abilities, and is accessible (provides vehicles with ramps), affordable, safe (during a day and night) and easy to use. It provides safe and comfortable places to wait, and stations with clean, barrier-free toilets, and stops are well lit and clearly signed with a voice countdown system and an emergency phone.

3   Understanding SMART City Concepts

The term "smart" became a direction in many urban agendas and policies of modern cities in order to direct the transition towards sustainable development goals. Smart city concepts are used all over the world with different interpretations, contents and values. Literature agrees that, generally speaking, a smart city means competitiveness and sustainability achieved by the use of information and communication technologies (ICT), and by development of human and social capital [6, 7]. Caragliu and Nijkamp highlight that a city can be defined as "smart" when *"investments in human and social capital and traditional (transport) and modern (ICT) communication infrastructure fuel sustainable development and a high quality of life, with a wise management of natural resources, through participatory action and engagement"* [8]. Smart Cities Council defines a smart city as *"one that has digital technology embedded across all city functions"* [9]. Singh identifies eight key aspects that define a smart city as *"smart governance, smart energy, smart building, smart mobility, smart infrastructure, smart technology, smart healthcare and smart citizen"* [10]. The Centre for Smart Cities writes that a smart city *"offers sustainability in terms of economic activities and employment opportunities to a wide section of its residents, regardless of their level of education, skills or income levels"* [11]. Jaššo and Petríková point out that a sustainable smart city is a *"meaningful network of links between physical environment, communication networks and social community"* [12]. Husár et al. on the other hand, consider the concept to be too vaguely defined arguing that the research on smart cities has been sparse and there is a shortage of critical analysis on what "smartness" means [13]. Similarly, some authors see the existing research to be more focused on the technical, economic and engineering dimension rather than social or political analyses [14].

Those interpretations generally sum up three main approaches to a smart city:

1. The high-tech-centred approach with a big emphasis on new technologies and ITC infrastructure that are key to achieve a smart city
2. The people-centred approach where a great deal of social and human capital determinates a smart city
3. The mixed approach that characterises a smart city as an integration of both elements above

We consider the third mixed approach, for a guideline of this chapter arguing that a city must achieve a healthy balance between technology and social capital in order to assure right conditions for sustainable urban development.

3.1   SMART Cities and People with Special Needs

All human beings possess a different range of abilities being able to do different things at different ages. Thus, an ability is a relative concept. When the ability relates to something considered as "normal behaviour", everything else (disability, handicap, age, etc.) becomes synonymous of abnormality. Mother with a baby in pushchairs, person carrying heavy loads, but also a young, healthy and fit man— almost everyone has experienced struggles in urban area at some point in his or her life. In this sense, we all can relate to some kind of disability. Wylde et al. assume that up to 90% of population becomes "urban disabled" in some way or another in their life [15]. According to the UN reports, 15% of the world's population lives with some kind of disability [1].

In the past, urban planning illustrated an urban user to be a young, physically fit, educated, middle-class (usually) male adult who embodied the anthropometric stereotype. The term originated with the birth of the Welfare State, and for majority of the twentieth century it was considered natural while any other conditions were ignored.

Against that, a term "environmental pressure" was used for a position of those who had to deal with environment built with regard to young, healthy and fit people. Later then, design for special needs, obviously, demonstrated requirements of all those people who did not fit the previous definition of an urban user—mentally and physically disabled, older people and women. With this perspective, people with physical and mental (sensory or cognitive) impairments became "people with special needs" [16].

Today, however, literature varies on what "special needs" means. Some explains that people with special needs are *"people who need special help or care, for example because they have a disability"* [17], and others by "special needs" describe a much wider community that includes:

- Physically disabled
- Mentally disabled
- Blind, visually impaired, low vision
- Deaf, hearing impaired, hard-of-hearing
- Elderly, seniors (include people suffering from disorders such as dementia, Alzheimer's, Parkinson's, poor vision, hearing impairment, balance problem, etc.)
- Homeless, shelter dependent (including shelters for abused women and children)
- Culturally isolated (includes people with little or no interaction outside of immediate community)
- Children, unattended minors
- Geographically isolated (with no access to services or information, immigrants)
- Poor, extremely low income, unemployed
- Single parents with no support systems
- Transient special needs—includes people temporarily classified as special needs due to a transient condition, status or illness (includes tourists as well) [18]

In this light, not only people with physical and mental impairments but also all above-mentioned categories can be considered as people with special needs. Each of the user category is characterised by its individualities, special needs and spatial requirements related to the urban environment and therefore any smart city needs to address those in order to achieve inclusive, accessible and sustainable urban development.

4 Design for All

Only until recently the special-needs approach was considered a reasonable solution for so-called architectural disabilities. Architectural disability is a term that describes a physical infrastructure (urban design) that treats people with barriers. The word "disability" relates to dysfunctional buildings (badly designed, built and poorly maintained) and enclosed, barrier and bounded places. Architectural disability makes the built environment inconvenient, uncomfortable and unsafe, and discourages people from using it [19, 20].

Despite good intentions, the problem was that the special-needs approach was based on needs not rights, which led to solving problems and not basic causal relations. Because of that, it was necessary to move from the rather fragmented design for people with special needs to "design for all" (also known as universal design or inclusive design). In 1997, a team of researches interpreted the "Seven Principles of Universal Design" [21]. The manual proposed design that *"enabled the equitable use of products and services by all, flexibility in their use, that products and services should be simple and intuitive, provide perceptible information, have a tolerance for error in use, require low physical effort and that designs should be at a size and allow for such space that would allow their use regardless of a user's body size, posture, or mobility"* [21]. Design for all aimed to create urban environment that would be usable by all without the need for "special design" adaptation. It described a design where all users despite their abilities could handle their everyday activities comfortably and safely. The main objective of design for all was to reduce environmental pressure and architectural disability, and to promote better social equity and justice [22].

4.1 SMART Design for All

Smart technologies have diffused into everyday life of people and significantly influenced urban settlements. They are supposed to improve the quality of life and have a strong influence on how people live and interact with each other, with their environment and public services. By 2025, demand for smart city services is expected to grow by more than 30% across Europe, Africa and Latin America [23]. However, the global transformation towards high dependency on technology may result in serious negative impacts on citizens, because if smart services are not accessible for all, they may deepen inequality, exclusion and isolation.

When we talk about smart technologies for people with special needs, cities need to be particularly attentive about their implementation and make sure that those smart technologies will not treat one's health and well-being. To become more inclusive and accessible, smart cities need to address smart services for people with special needs across different areas: housing, transportation system, healthcare, public participation and citizens' engagement, community support services, public services, public space, leisure and culture. Smart cities should build on the principles of design for all and when implementing smart services move their urban agendas towards SMART design for all. SMART design for all links the core principles of design for all and the use of smart technologies in order to provide better quality of life for all, particularly people with special needs.

People with special needs most often meet with barriers of mobility limitations and visual and hearing impairments and they are highly sensitive of their urban environment. Smart technologies might help overcome mobility, visual and cognitive problems. Below we summarise some of the current technologies that are addressing problems of people with special needs in urban areas. Examples of smart technologies for:

1. Visual problems:

   (a) Perceptible (audible and vibrational) signals for pedestrian expanded to systems capable of telling people where they are
   (b) Accessible shopping for visually impaired people via mobile technologies (navigation system and a product recognition system)
   (c) Assisted city applications fit to blind users

2. Hearing problems:

   (a) Systems which convert voice to text or translate sign language

3. Cognitive problems:

   (a) Gadgets that are able to guide elderly people on their everyday tasks
   (b) Rehabilitation systems and video games to boost cognitive functions

4. Mobility problems, problems with navigation:

   (a) A mobile technology that provides audio instructions to visually impaired citizens who travel by means of public transportation
   (b) A smart parking technology that uses sensors by which city authorities can monitor parking lots and inform the disabled about available parking spots so that they can find it right at the time

5. Personal and home accessibility:

   (a) Systems for regulating the heating, air conditioning, lighting and water temperature via smartphones
   (b) Products that enable to control lighting systems, blinds and doors by tablets or key fobs or a voice recognition technology
   (c) Gadgets that can address requests in voice language and ask to play music, read an audio book, order a ride or groceries
   (d) A smartwatch with an emergency button to contact a live support team and get medical help 24/7

Using smart sensors, navigation solutions and modern communication technologies, people with special needs have a chance to live more independently and become more engaged citizens. Apart from those, SMART design for all can also facilitate better social interaction and social activities. Studies have proved that social interaction helps seniors as well as disabled to keep good emotional and physical health together with good cognitive functions [24]. Smart technologies can help the elderly and disabled people maintain social interactions with families, friends and close communities; actively participate in their environment; and exchange new experience with others.

In conclusion, smart cities need to develop complex strategies on how to make mainstream and smart services accessible for all and how to use smart technologies and services to serve for people with disabilities, seniors and other disadvantaged people to improve their social life, safety, healthcare and participation in urban life. If cities succeed to promote smart information to all citizens, SMART design for all has a high chance to become a part of everyday life.

5   Critical Discussion

The quality of public spaces significantly influences the quality of people's lives. It has been widely researched that inclusive urban planning has a significant role in strengthening social inclusion [25]. But even in the modern world led by smart technologies, urban inequalities are still being "built" into new places. The reason why urban planning has been constantly failing to meet challenges of increasingly diversified society may stem from the fact that historically the attention was paid to design which would suit "normal" people. The response to that was the effort to design for "special needs" which again led to the separation of the mainstream society and minorities. Many critics also argue that a disability is first and foremost a prejudice invented and nourished by a society that dispraises disadvantaged people [26]. This assumes that a person becomes disabled by the barriers they must face, not by their impairment. Until such basic issues survive in modern cities, smart concepts remain nothing more but empty words.

6   Conclusion

Contrary to expectation, we are at a point where the idea of smart cities is still developing and the process of implementing smart solutions is still in progress. As we researched in the first part, there are different definitions of smart city concepts and what "smart" means. New technologies are clearly the cornerstone of smart cities worldwidely; however by "smart" we understand cities that are primarily inclusive and accessible for all. The smart city is not a product but rather a complex process of city transformation which needs to reflect specific needs and requirements of its citizens. The attention in many recent definitions of smart city has also switched towards more "human factor"-related issues.

We have recognised the need for more inclusive and accessible smart city models that would emphasise the importance of social inclusion and social cohesion in sustainable urban development policies and agendas. We call for a greater focus on different social groups, particularly the most vulnerable of them—people with special needs, who have always been overlooked in mainstream policies. We believe that inclusive and accessible smart cities will be able to create space for the acceptance of those specific groups. However, this requires both citizens' engagement and new ways of cooperative and collaborative approach in urban-related issues.

Including the needs of specific social groups in urban planning moves the concept of smart cities much closer to the concept of smart cities for all. SMART design for all, inclusivity, social interaction and accessibility are some of the key aspects of this concept. The equality in terms of urban planning is essential for a smart city that directs its development towards more sustainable urban future.

It is clear that smart cities won't happen overnight and therefore it is time to rethink the current development towards more integrated model where people with special needs are well communicated and addressed in both global and local smart city policies.

Acknowledgements The research leading to these results was conducted in the frame of the project "Socio-economic and Political Responses to Regional Polarisation in Central and Eastern Europe" (RegPol²). The project received funding from the People Programme (Marie Curie Actions) of the European Union's Seventh Framework Programme FP7/2007-2013/ under REA grant agreement no. 607022 and national grant scheme Vega 2/0013/17 and project SPECTRA+ No. 26240120002 "Centre of Excellence for the Development of Settlement Infrastructure of Knowledge Economy" supported by the Research and Development Operational Programme funded by the ERDF.

References

1. WHO, World Report on Disability, ©World Health Organization 2011, 2011 [Online]. Available http://apps.who.int/iris/bitstream/handle/10665/70670/WHO_NMH_VIP_11.01_eng.pdf;jses sionid=57645F2039C9CB654FB416189D575A99?sequence=1. Accessed 22 June 2018
2. European Commission, Population ageing in Europe, General for Research and Innovation, 2014 [Online]. Available https://ec.europa.eu/research/social-sciences/pdf/policy_reviews/kina26426enc.pdf. Accessed 22 June 2018
3. D. Petríková, S. Ondrejičková, Quality of space in cities respecting requirements of specific target groups as objective of smart city concepts, in: *Smart City 360°*, vol. The second EAI International Summit, ISBN: 978-1-63190-149-2 (2016)
4. M. Finka, L. Jamečný, V. Ondrejička, Urban safety as spatial quality in smart cities. Smart City 360°, 821–829 (2016). https://doi.org/10.1007/978-3-319-33681-7_73
5. Commission for Architecture and the Built Environment, Inclusion by design: equality, diversity and the built environment, Commission for Architecture and the Built Environment, 2008 [Online]. Available https://www.designcouncil.org.uk/sites/default/files/asset/document/inclusion-by-design.pdf. Accessed 22 June 2018
6. R. Giffinger, C.H. Fertner, H. Kramar, R. Kalasek, N. Pichler-Milanovic, E. Meijers, *Smart Cities—Ranking of European Medium-Sized Cities* (Vienna, Centre of Regional Science, Vienna UT, 2007)

7. E.P. Trindade, M.P.F. Hinnig, E.M. Da Costa, J.S. Marques, R.C. Bastos, T. Yigitcanlar, Sustainable development of smart cities: a systematic review of the literature. J. Open Innov. Technol. Market Complex. **3**, 11 (2017). https://doi.org/10.1186/s40852-017-0063-2

8. A. Caragliu, C. Del Bo, P. Nijkamp, Smart cities in Europe. J. Urban Technol. **18**, 65–82 (2009)

9. Smart Cities Council, Definitions and overviews, SCC, 2016 [Online]. Available http://smart-citiescouncil.com/smart-cities-information-center/definitions-and-overviews. Accessed 22 June 2018

10. S. Singh, "Smart cities—a $1.5 trillion market opportunity", Forbes, 2014 [Online]. Available http://www.forbes.com/sites/sarwantsingh/2014/06/19/smart-cities-a-1-5-trillion-market-opportunity/#2a57763f7ef9 Accessed 22 June 2018

11. Center For Smart Cities, Draft concept note on smart city scheme, Government of India, 2014 [Online]. Available http://c-smart.in/wp-content/uploads/2015/02/CONCEPT_NOTE_-3.12.2014__REVISED_AND_LATEST_.pdf. Accessed 22 June 2018

12. M. Jaššo, D. Petríková, Towards creating place attachment and social communities in the SMART cities, in *Smart City 360°: First EAI International Summit*, October 13–16, Bratislava, Slovakia and Toronto, Canada, Smart City 360°, 2016

13. M. Husár, V. Ondrejička, S. Ceren Varis, Smart cities and the idea of smartness in urban development—a critical review, in *IOP Conference Series: Materials Science and Engineering*, doi: https://doi.org/10.1088/1757-899X/245/8/082008, 2017

14. S. Marvin, A. Luque-Ayala, C. Mcfarlane, *Smart Urbanism: Utopian Vision or False Dawn?* (Routledge, London, 2015)

15. M. Wylde, A. Baron-Robbins, S. Clark, *Building for a Lifetime; The Design and Construction of Fully Accessible Homes* (Taunton Press, Newtown, 1994)

16. A. Sklar, J. Suri, Us vs. them: designing for all of us, in *Proceedings of Include 2001*, 18–20 (2001)

17. Collins English Dictionary, Definition of 'Special Needs' (Harper Collins Publishers, 2018) [Online]. Available https://www.collinsdictionary.com/dictionary/english/special-needs. Accessed 22 June 2018

18. Collaborating Agencies Responding to Disasters, What do we mean when we say "people with special needs"? in: *Serving People with Special Needs in Times of Disaster*, April 2004 [Online]. Available http://emergencypreparedness.cce.cornell.edu/family/Documents/PDFs/CARDspecialneeds.pdf. Accessed 22 June 2018

19. W.S. Atkins, *Older People: Their Transport Needs and Requirements* (Department of Transport Local Government and the Regions, London, 2001)

20. G. Selwyn, *Designing for the Disabled*, 3rd edn. (Routledge, London, 1976), p. 145

21. B.R. Connell, M. Jones, R. Mace, J. Mueller, A. Mullick, E. Ostroff, J. Sanford, E. Steinfeld, M. Story, G. Vanderheiden, Seven principles of universal design, in: *The Principles of Universal Design*, North Carolina, The National Institute on Disability and Rehabilitation Research, U.S. Department of Education (1997)

22. Center for Universal Design, Principles of Universal Design, North Carolina State University: Center for Universal Design (1995)

23. The Global Initiative for Inclusive Information and Communication Technologies, Smart cities & digital inclusion, G3ict and World Enabled [Online]. Available http://g3ict.org/design/js/tinymce/filemanager/userfiles/File/Smart%20Cities/Smart%20Cities%20and%20Digital%20Inclusion-concept%20note.pdf. Accessed 22 June 2018

24. J. Barnett, K. Vasileiou, F. Djemil, L. Brooks, T. Young, Understanding innovators' experiences of barriers and facilitators in implementation and diffusion of healthcare service innovations: a qualitative study. BMC Health Serv. Res. **11**, 342 (2011). https://doi.org/10.1186/1472-6963-11-342

25. F. Schreiber, A. Carius, The inclusive city: urban planning for diversity and social cohesion, in State of the World Book Series: Can a City Be Sustainable?, (Washington, 2016), pp. 317–335

26. A. Mahmoudi, M. Mazloomi, Urban spaces, disabled, and the aim of a city for all: a case study of Tehran. Int. J. Sci. Basic Appl. Res. **1**, 530–537 (2014)

Chapter 6
AI, IoT, Big Data, and Technologies in Digital Economy with Blockchain at Sustainable Work Satisfaction to Smart Mankind: Access to 6th Dimension of Human Rights

Andrea Romaoli Garcia

Abbreviations

AI	Artificial intelligence
BEPS	Base erosion and profit shifting
BRL	Brazilian real
CAC	Cyberspace administration of China
DNS	Domain name system
DLT	Distributed ledger technology
GDP	Gross national product
GFI	Global financial integrity
ICANN	Corporation for assigned names and numbers
ICO	Initial coin offering
ICT	Information and communications technology
IOF	Tax on financial operation/transaction
IoT	Internet of Things
IP	Internet protocol
IR	Income tax
ITCMD	Tax on the transmission of assets by death or donation
ITU	International telecommunication union
MRTI	Matrix rule of tax incidence
OECD	Organisation for Economic Co-operation and Development
OIT/ILO	International labour organization
TLD	Top-level domain

A. R. Garcia (✉)
Chicago, IL, USA

UNITED NATIONS. IBET UNIVERSITY. INTERNET SOCIETY-NY. ICANN.WTO.
ITU-UN., UNITED NATIONS., Geneve, Switzerland

© Springer Nature Switzerland AG 2020   83
N. V. M. Lopes (ed.), *Smart Governance for Cities: Perspectives and Experiences*,
EAI/Springer Innovations in Communication and Computing,
https://doi.org/10.1007/978-3-030-22070-9_6

UN   United Nations
UN/DESA   Department of Economic and Social Affairs - United Nations
UNCTAD   United Nations conference on trade and development
UNESCO   United Nations educational, scientific and cultural organization
USA   United States of America
WBG   World Bank group
WFP   World food program

1 Introduction

This chapter is the result of 2 years of researches about technologies and interactions with the humanity in social and financial aspects and the author was asking to himself: What impact the disruptive technologies are bringing to worldwide economy and why are artificial intelligence, IoT, big data, and blockchain considered disruptive?

The research went beyond the answers it sought to solve stormy issues typical to this century: there is a broad and effective idea conceptual behind the disruptive technologies.

Throughout history, five dimensions of basic needs have been created to establish the existential minimum.

This research demonstrated that fundamental rights reached the sixth dimension of human rights in 2009 where the right of access to technology is fundamental to generating peace, dignity, and a sustainable world. Establishing technology as a fundamental right is useful to guide the Annual Budget Plan by the rulers who use the fundamental rights to establish the margin of investment to be made in each sector proportional to the importance level as it means for the citizen's life.

For these reasons, the research brought evidences that blockchain and cryptocurrencies are much more than a payment system. The method used was scientific with analysis of historical documents and social observation also well seeking laws, doctrine, and jurisprudence confronted with data taken from socioeconomic statistical reports collected in 2017.

This research was enough to show that the blockchain technology and cryptocurrencies combined with AI and IoT compose the smart economy. Through poverty reduction, physical borders will be protected by virtual borders. In the end, the author concludes that technologies mean a powerful accelerator of world economic flow and it means a vehicle for the sustainable world requiring for investments like a human right of sixth dimension that can't be separated off another rights.

It was analyzed like a social-economic event that is modifying the traditional financial system since the blockchain technology, smart contracts, and cryptocurrencies came up.

Also, the International Tax Law scenario was examined by side ideals from governments in a democratic system as an instrument that materializes human rights.

On the other hand, the author has brought elements to affirm that high taxes jeopardize the future of mankind and that less greedy taxes bring greater profit and growth to governments.

Philosophy was studied as a scientific way to explain the reason that technologies are causing a boiling in all layers of society and in all countries at the same time. Through philosophical and sociological studies, it was possible to bring evidences to raise technology to the sixth dimension of human rights.

The social observation approaches the legacy of Emile Durkheim philosophy who established a power relationship between social fact and coercion. The taxation has been the focal point in smart economy for juridical scientists and everyone involved in the digital economy. Some countries made a regulatory framework but there is still no international inclusive framework. For this reason, the premise for legislation on disruptive innovation is a modern legal theory to serve as a basis for validity of legal norms in smart economy that was the object of study. This research conducted testes and researches from historical and social method, seeking for laws, doctrine, jurisprudence, and concrete case analysis in front of the philosophical school of logical-semantic constructivism as a suitable means to verify the possibility of constructing a matrix rule of tax incidence. This research was enough to show that the blockchain technology and cryptocurrencies fulfil its humanitarian role in the smart economy.

When we put technology into the sixth dimension of human rights it brings up a complete legal doctrine useful to rulers and decision makers. Thus, it drives the technologies further.

The author created the full and strong legal doctrine so that other lawyers and judges can base their decisions because we cannot forget that this has an impact on people's lives.

The disruptive technology that includes artificial intelligence, IoT, big data, blockchain, and smart contracts deals with people's data, life, health, and money and consequently some cases will be subject to review in international courts. Lawyers, attorneys, and judges need grounds for decisions.

The technology, especially artificial intelligence, IoT, big data, smart contracts as the sixth dimension of human rights, financial system that is inserted, cryptocurrencies, as well as impact that taxation brings to humanity, has already been presented at the IGF Forum 2018 at UNESCO [1] Headquarters, Paris, for the technical and scientific community from Internet governance and regulatory framework. In the same way, it was presented at the conference for technology WEBbr 2018 [2] in Brazil and was spread by Elon University from the survey about artificial intelligence and future of humans. From the United States, Elon University [3] is recognized by the National Survey of Student Engagement as one of the most effective universities in the nation.

2   The Social Fact and the Constructive Process

Until the year 1800, the world was almost homogeneous: most people were poor and the range of professions was restricted.

Thus, the French Revolution (1789–1799) becomes a milestone for our analysis about the human rights evolution embodying the importance of technology to humanity since it was a disruptive period socially, economically, and politically.

Civil society was divided between the clergy, the nobility, and the bourgeoisie. The bourgeoisie was the part of society that paid taxes that were used to pay for the good life of the court, clergy, and nobility.

One special fact is added; the country went bankrupt and the blame was attributed to the poor economic administration of the king who also controlled the courts appointing sentences with the content almost always unjust and impartial.

The bourgeoisie joined the poor classes because the poor economic management of the king didn't provide the growth of capitalism. Thus, the greater freedom for trade and the end of high taxes were, among others, the most important causes of the French Revolution.

Thus, by ROUSSEAU, LIBERTÉ, EGALITÉ, FRATERNITÉ [4] was the slogan of the French Revolution and marked the end of feudal privileges, the equality of all before the law, and the guarantee of property. THE DECLARATION OF THE RIGHTS OF MAN was approved by the French National Assembly on August 26, 1789, which decided that a declaration of rights should precede the Constitution.

The French Revolution meant the end of the absolutist system and the privileges of the nobility. The people gained more autonomy and social rights came to be respected. In the meantime, the bourgeoisie conducted the process to ensure its social dominance.

Is possible to see that autonomy and social rights always was walking by the side of human evolution. The more the evolution, the more the autonomy to society and the less the interference from state in individual life.

For many centuries, the coins that were references to each country didn't generate discussions about legality or doubts to insert the way to taxation. There were different names to refer a commodity, generally gold or silver, applied monetarily. This universalism over the money was useful to all countries because it promoted free trade; it was aiding merchants in economic calculation, and represented a solid and reliable means of concentrating power with rulers.

By the way, the governments used the practice of depreciating money to maintain and increase state power because it was a faster and less critical means than the traditional method of taxing the population to give power to the public coffers.

It means an inefficient economic policies when governments are using a method prioritizing the power and coercion over money to leverage the economy despite it increases the living expenses that it is financed by the gradual impoverishment of the population to do the balance in the monetary system.

Since the World War I (1914–1918), the method adopted was the issuing of paper money to make revenue and financing the costs of war that led the world into widespread bankruptcy. After the 1970s and till 2008, currency reform plans spread throughout the world in forums and discussion groups aimed at strengthening the world economy and reducing poverty. However, governments were not interested in this discussion that threatened the monopoly of power.

But in 2008, Satoshi Nakamoto, called as "cyberpunk," took the incredible initiative of reinventing the currency in the form of computer code. The result introduced Bitcoin to the world. Nakamoto released it with a white paper in an open forum saying simply: "Here is a new currency and a payment system. You can use it if you want."

Bitcoin was absolutely non-reproducible and constructed in such a way that its historical record of transactions made it possible for each monetary unit to be reconciled and verified in the course of currency evolution.

The digital economy came up enlarging the concept about democratic legitimacy to drive the private life through the financial autonomy. The payment system with cryptocurrency occurs without higher bodies establishing the value or authorizing the legal owner in transactions. Thus, the ecosystem is an open source network that it removes obscures points of failure. Encryption, a distributed network, and a continuous development that is made possible through developers paid for by the transaction verification services provided entered into the system. Those combined attributes created and reinvented money in an interesting system to users and without interference from governments or banks. This meant pure materialization of a noble democratic principle about legitimacy: citizens acquiring financial autonomy through the less intervention from higher bodies on private life.

This new age open the NEO- CONSTRUTIVIST DEMOCRACY emerging from the Hyperconnected Global Community as well creating a new consumerist society, development strategies and business intermediation in the global economy by making feasible the use of tools or utilities to expand the manufacture process, increase knowledge, and allowing the accomplishment of multilateral business through computer communication (internet).

The B2B was reinvented by the neo-construtivist democracy.

Actually it is an emerging international currency created by market forces with a foundation in entrepreneurship, market exchange (barter), and democratic principles. By the way, to ensure the focus about our main subject studied here, the authors didn't bring the public and uncontested data table on the quantitative evolution of the currency.

After study, the author realized the social fact as the trigger event of this new currency and it is being drawn by the functionalist theory and collective consciousness of social fact as studied by Emile Durkheim (1858–1917), where it is possible to know that the society is drawn by the role that the individual develops in the group and the forces that weigh on him/her will make him/her react in self-defense (REALE, 1998).

Durkheim's studies led him to state three attributes for the social fact: generality, exteriority, and coercivity. The generality refers to the collective social facts that affect the whole society (collective). The exteriority means the characteristic of the social facts existing when the person is born and this is external to this person who hasn't got the mastery of choice over them and must react as they are presented. Coercivity is related to the power or the force that the patterns of culture are imposed on the members of a respective society. This characteristic pushes the individuals to meet cultural standards.

In this way, the collective consciousness comes with the socialization of the individual provided by the environment and culture that this person is inserted.

The emergence of cryptocurrency comes as a response to the pressure that the individual has been suffering since the end of World War I that was responsible for global impoverishment to generate security and enrichment for the governments and banks. About banks, they were enriching themselves taking advantage from people that were needing financial assistance.

A situation of high state interference was being drawn in the individual private life and that is incompatible with the democratic system and collective consciousness.

The return to the teachings of the social fact as Durkheim taught us is important now because it represents a logical way for the new message created by blockchain technology and the cryptocurrencies to reach interested parties and be part of positive law as a tax legal fact. The technique to interpret this fact goes through the semantic logical constructivism theory.

The juridical study is starting from the point the right language will be built through the facts that arise and from which are extracted the necessary signs to construct the message through signification.

In respect to all philosophical schools that aim to conceptualize the law science and validate norms, these theories don't explain the new socioeconomic event because it isn't exhausted in the concept of natural law and lacks subsidies to ground that just the justice and validity of the norms would be coincident. For our study, these isolated concepts will not be useful as a premise to validate the financial system in the context of blockchain technology and virtual currencies.

After these considerations, there is only possibility of transcribing this event to a competent legal language aimed to validate and support legality the institution of taxes through the philosophy contained in the theory of semantic logical constructivism. This technique allows the possibility of dialogues with other methods and it will not commit the sin of studying the object in isolation without the methodological cut necessary and appropriate to reality.

Talking about philosophy of language, to make right and logical conclusion to claims if it will be possible to increase the activity inside blockchain with taxes, also the authors will consider the linguistic turn knowing that it isn't ground on absolute truth and it will be a human responsibility to do the interpretation to build the sense. The act of interpreting is based on language and will create reality through associations starting from a referential point such as time, place, space, and culture of the cognoscente subject, based on a model. The logical-semantic constructivism is an instrument that adjusts the form and substance from thought to join the means and process to construct the legal text in view that language is always part of legal science.

It brings logical schemas to our thoughts materializing the language with a small degree of certainty by combining the syntax of the words and the content throughout choices most appropriate to the event that presents itself. These signs of communication are studied by semiotics and they are on the side of logic. The next step is to add the semantic load to find to know the truth about the event (CARVALHO, 2014).

In this way, the cryptocurrency event represents the communication between the citizen of a specific country, the government, nongovernmental organizations, companies, and rules with some degree of coercivity referring to the traditional financial system when compared to the new innovator system in the field of virtual communication technology.

At this point, the term "virtual" requires a few notes. When it is said that the currency is virtual it doesn't mean "unreal" coin. The Bitcoin currency and other correlates are a real coin, and virtual is the technology employed to bring them to reality.

Law science as a way to communications has the responsibility to build meaning and understanding of the text to draw the reflexes that the blockchain brings to society, governments, and international tax law. This way the law science will be able to tell a new reality. The currency in question is cross-border and only deals with "legal compliance for each country respectively." These are the only rules that the new chain of wealth, jobs, and taxes needs to deal.

The referential model as a determining factor that is accepting the truth as relative doesn't means denying the true propositions, because the truth is relative and not absolute since the interpreter will seek to attribute meaning to the fact by creating propositions.

Along the text, the matrix rule of tax incidence (MRTI) will be built just in case it is possible after all points are analyzed. But to come there, there is a necessity to know this possibility in details about the scenario in digital economy with artificial intelligence, disruptive technologies like blockchain, and cryptocurrencies being the social event as a relativity to build legal propositions. In the end, we can know if governments will be able to establish taxes and rules and how they will establish the regulatory framework.

So, if this is the premise adopted here in this study, we can't deny the impact that the new reality in blockchain's technology brought us since society responded to what existed in the traditional oppressive financial system with determination and effectiveness. The marks of this social reaction are bringing social and economic impacts. Behold, popular reactivity can't be neglected since it has added relevant technological development.

That said, the governing authorities have no other way to create a minimum protective rule appropriate to the function that the virtual currencies are playing.

From a lawful view event and fact means differents concepts and signs.

Event is the presupposition for fact, that is, the event is the occurrence in the fact's world that has not been converted into a linguistic account. On the other hand, fact is the event that supposedly happened and concerns the denotative statement about a situation considered in time and space. This will always be referred to the past. The fact becomes legal when pronounced through legal speech.

Forward, the RMIT refers to the tax law in the strict sense, since its core is essentially the definition of a general and abstract legal standard capable of identifying the minimum elements of the tax standard. It is hypothetical-conditional judgment that embodies a hypothesis and a consequence.

In the development of RMIT we establish elements to determine the antecedent and the consequent. The antecedent is determined by three criteria: material, spatial, and temporal. The consequent shows us two criteria, personnel and quantitative, and it is determined by the calculation base or assessment basis and aliquots.

In summary, the RMIT hypothesis picks up an event forecast, and when this is concretized the interpreter will describe in technical legal language establishing the obligatory link between the taxpayer (taxable person/duty to perform the tax liability) and the state (active subject/will receive the tax liability).

There is a legislative vacuum for the economic and financial system of investments in the blockchain field and the use of virtual currencies. We note that this gap exists throughout the world.

After these considerations, it isn't possible to deny the tax effects and the need to bring a minimum rule for this field. There is no compassion for ratchet effect because the social retrocession is impossible to admit it within the context of legal certainty.

Although it is implicit and fruit of doctrinal creation, the principle of the social setback is the necessary consequence from superior principle of legal certainty.

This constitutes the bond to the protection and progress of fundamental human rights that legitimize constitutional guarantees to taxpayers. The predictability in intersubjective and taxpayer relations with the treasury prevents regression.

Establishing standards in this field means making rules with minimum interference in their operating environment under the risk of mischaracterizing the function and purpose of the virtual currencies. We must remember that this coin has got its origin in the absence of state control and they occur outside the banking institutions. Also, they use only the electronic communication of technology in peer-to-peer chains by addressing the domain name system (DNS) and ICANN.

This way, the taxpayer must declare spontaneously and they will get under a incidence of Income Tax if there is a capital increase resulting from in disposal of assets. The problem arises from a absence the tax legal nexus between tangible and intangible goods and services to point out the legal permanent establishment to define which country will demand for taxes.

This nodes makes companies is under double taxation and citizens don't receive the benefits from a EQUITABLE JUSTICE. The OECD defines the double taxation as an unlawful practice because it will bring grave impact to economy making taxes working as barriers to growin of the World Economy Flow. The OECD Model Tax Convention, a model for countries concluding bilateral tax conventions, plays a crucial role in removing tax related barriers to cross border trade and investment. It is the basis for negotiation and application of bilateral tax treaties between countries, designed to assist business while helping to prevent tax evasion and avoidance.

However, in the international scenario this assumes diverse features and it justifies the reason why the legal framework is requiring specific approach considering case by case. The cryptocurrencies, blockchain technology, artificial intelligence, and all disruptive technologies are already a reality and evolve every day demanding a behavior typical from legal professionals, governments, and economists. The new social and economic reality required the quest to interpret reality, observing the events and their signs and applying the accumulated knowledge.

Since the declaration of the rights of man, many constitutions have been elaborated around the world having democracy as a model of government. This was the first bill of rights and source of inspiration to the United Nation universal Declaration of Human Rights in 1948.

However, as the authors analyze, since the end of World War II (1939–1945) the states have developed mechanisms to regain privileges, control the flow of economic growth, and interfere with property rights in confiscatory acts that, although legal, are immoral and unconstitutional.

Throughout the text, the authors develop the theme to conclude that human rights are currently demanding the sixth layer of protection for the rights claimed since the French Revolution: the right to technology as a form of financial independence to provide the minimum existential life worthy of form to ensure the survival and development of future generations in all respects in a healthy environment that is available to all.

This is easy to verify. Broadband access is an aspect of technology that causes changes in the life of the individual that when accessing the Internet some things will come up. People will become more proactive and experience the satisfaction of achieving positive results for their lives. When they have access to the Internet, they can see how the world works. It means the sustainable work satisfaction.

The World Bank Group (WBG) in April 2018, after studies, said that technology and economics go hand in hand to allow for decent living as reference income increases and individual income, proportionately, also increases.

Similarly, partnerships between civil society, private companies, governments, and nongovernmental organizations can improve the economy and facilitate the technological education in developing countries on a focal way and faster than governmental actions alone.

Technology will do us a great service by connecting everyone, but the other thing that technology is doing at the same time is to modify the way in which jobs claim larger investments than those made in the framework of laws to increase the power of state administration through increase of taxes.

Raising taxes as a way of generating wealth and reducing poverty is an outdated and insufficient mechanism that does not guarantee sustainability.

The 6th dimension of human rights is on the center of the Fourth Industrial Revolution or the "disruptive technological age" since it will allow interactions between humans, machine, and robotics that are able to collaborate across sectors, borders, and fields to grasp the opportunities it presents.

3   Fundamental Notes for an Analytical Positioning

The double taxation, high taxes, and the state's seeking for profit from taxes prevent the positive advance of the global economic flow and the technological innovation.

Alongside, the digital economy is one of the goals for the 2030 Agenda for Sustainable Development from Department of Economic and Social Affairs in the United Nations because there is no doubt that it has the potential to accelerate economic and educational development as well as provide a protected environmental for future generations.

However, we don't have an inclusive standard till now for regulating the digital economy with blockchain and cryptocurrency and establishing high taxes will create risks to technological innovation.

Watching the landscape, there arise three problems to humanity survival:

1. Increasing the economy and creating jobs is an urgent necessity and it must be done now and fast.
2. High taxes are jeopardizing the human survival and the double taxation cannot be admitted because this represents illegal confiscation on tax base.
3. Governments must manage the state activities in the same time that they assume the responsibility for human security in a global field.

This research provided a legal doctrine that will guide the regulatory framework as an inclusive global standard and also will drive the governments, lawyers, and judges to maximize the applications of taxes.

Nowadays, the Bitcoin means the emerging international currency, created entirely by market forces. It must be argued that against market forces governments can influence nothing in order to restrain, modify, or appropriate.

Bitcoin is the first fully decentralized global payment system as well as it is a money like the real, the dollar, or the euro in digital method; it is not issued by any government and its value is determined freely by the individuals in the market. It is the reason why Bitcoin is a totally revolutionary financial system. The transactions to payment are cheaper and more secure than classic method from banks. With Bitcoin you can transfer funds from A to B anywhere in the world without ever having to rely on a third party for this simple task. Technically, Bitcoin is a digital currency peer to peer, open source, which doesn't depend on a central authority. Although at first glance it may seem complicated, the fundamental concepts are not difficult to understand. A monetary unit of Bitcoin is simply an electronic accounting note where the current account (the Bitcoin address or the public key) and the balance of Bitcoins at a given time are recorded. In this sense, a unit of Bitcoin doesn't differ in anything from a unit of real or dollar deposited in a bank, since it is also a mere electronic accounting record. But there is a big difference; in the case of Bitcoin, the space in which the records are concluded is unique, universal, and shared by all users (the blockchain), while in the current system each bank owns and controls its transaction log (its own ledger). This difference assures greater accountability to the digital payment system, ULRICH [5].

Indeed, the cryptocurrencies means more than Bitcoin. The digital market is working with another like Ethereum, Litecoin, Tron, and others.

The financial system is being reinvented against willingness from bankers and this didn't come about because of international conferences or because an academic group joined to formulate a plan. But this social event, subject of our research, is justified because science and technology reactively join cyberpunks in the name of a worldwide need: to acquire greater financial privacy, access to wealth, access to jobs, financial freedom and power in conducting globalized business without the interference of payment intermediaries, and less fees and costs in payment transactions.

Blockchain came up renewing all markets, minds, and life. Therefore, blockchain is an inspirational software embodying unique attributes. As an example, the crowdfinding allows the interaction between blockchain, smart contract, community, private companies, and governments.

The legal concept about crowdfunding mechanism is established by the author, ROMAOLI, like the raising funds from small groups and it aims to finance the building schools, universities, or hospitals for the maintenance of social services of governments. This mechanism has much to get benefits from the 6th dimension of human rights theory because it is a collaborative economic activity to the government and through the decentralized public administration. The governments can collaborate on these projects by indirect public administration. It is just one among other ways to maximize the application of taxes. However, in order to allocate resources, the government must respect the annual budget law that distributes portions of financial resources according to the need of the activity in question. It is a formal law that cannot make changes to the legislation in force and cannot introduce new obligations.

Thus, the blockchain's financial system runs from the outside-in and from the bottom-up, based on the principles of entrepreneurship and market exchanges [6].

In fact, what happens inside blockchain is the genuine data exchange that results in financial value only at the end.

From these premises, it is possible to affirm that the blockchain technology allied to Bitcoin is a system that will face problems typical of innovation. However, it is a potential that will ascend into the future and it doesn't back down.

Bitcoin has inherited all the best characteristics existing in the best money: scarce, divisible, high portability, incorporeal, no intermediaries, solid, not under the power of government intervention, and universal. These are characteristics that facilitate the transfer of ownership at zero operating cost. Transaction rates are practically zero. In addition, there is security, fraud protection, speed, privacy, and more that make Bitcoin a superior technology that is requiring investments in this field.

This new financial system with Bitcoin isn't perfect now and there is high risks. But it is requiring just a few adjustments. So, some rules or government standards will fix it and make it in a better option as a financial system. In this way, it will manage the situation aggressively around the whole banking system that suffers the interference by the power of governmental intervention that does not always move to meet the public interest.

It is worth mentioning, however, that some governments don't accept the digital economy because they are feeling the power under threat, despite being powerless against market forces.

However, there is a loophole that allows rulers authorities to establish tax through the currency conversion operation; this will be possible when it is exchanged for another traditional currency that circulates with the intermediation of a financial institution. It is the moment that the material, spatial, and temporal requirements from tax matrix rule are complete and the tax incidence factor would be authorized. Even so, it is possible to apply the tax exception for reasons that will be detailed along the text.

The taxes are a node requiring political wills to solve and impact investments to bring a security environment to smart economy. By the way, it is possible to apply some solutions to the regulatory framework that serves for legal defense of block-chain for goods.

Another point is how the semantics is an important aspect about linguistic to building and interpreting the law when the need arises to make the regulatory framework for new social-economic events as in the case of disruptive theory.

The construction on legal theses requires a certain degree of pragmatism, practical results and in line with reality. Thus, it is highlighting the Blockchain's technology since it emerges bringing security and accountability to transactions with Bitcoin fixing the problem of "double spending" by preventing duplicate payments and solving an old problem. By way, we already have other cryptocurrencies to payment system and investments into digital companies that is showing broad performance such as ETHERUM.

Combining the result from these researches with the great and invaluable technical contribution brought by the expert Ulrich [5] about this technology, I will describe its functioning and interactions.

When we think about blockchain technology we realize that this means the genesis block that when encoded in the software serves as the starting point of the system loading information about the rules or instructions about the remaining database.

In a subsequent step, the database is formed from a series of blocks which together form a chain. This is where the name "block chain" or public transaction record (blockchain) comes up. Each block in the chain contains information or transactions. This is the system that solves the problem of "double spending" and without the need for intervention by a third party. Bitcoin does this by distributing the historical record to all users of the system via a peer-to-peer network, or I can simply say delivery of the data (message) to the recipient in paired register.

All transactions concluded in the Bitcoin economy are recorded in a kind of public ledger and distributed in blocks called blockchain. The ledger is the mains grouping of accounting records of a company that uses the double-entry bookkeeping method or standard record system. As you add transactions, your information is stored in the block according to when it was processed.

After the transactions are piled in the block, a hash is added at the end of the block. The hash is connected to the previous block of the chain. These hashes form the bonds returning between the chains until they reach the genesis block. The hash includes the current block number and the next block number in the chain. It also includes the date and time that was signed in addition to the amount of transactions included in the present block. The hash appears as an encrypted key and cannot be modified and it is impossible to recreate the input value. Encryption is algorithms that carry encryption for those who have the access key.

The computational force performs the records and the reconciliations of the transactions under the responsibility of the users. These users, they can be miners, are rewarded for their work with newly created Bitcoins. The Bitcoins are created, or "mined," as thousands of scattered computers solve complex mathematical problems that verify transactions in blockchain.

In addition to everything else, this system brings new professional positions that it isn't placed in a new list because they were nonexistent until digital economy entered the world economy.

How is currency created in the virtual world to be used in the real world?

The answer is science and technology. What happens in the case of Bitcoin is a search through computers to find the correct data sequence (the "block") that produces a certain pattern when the Bitcoin's algorithm "hash" is applied to the data. When a combination happens, the miner earns a Bitcoins award (and also a service fee, in Bitcoins, if the same block was used to verify a transaction). The size of the prize is reduced while Bitcoins are mined.

The term "mining" wasn't adopted by chance and concerns the mining process of Bitcoin which was a technology designed to recreate the process of mining gold and precious stones in the virtual network. And just as mining takes place in the material world, Bitcoin mining is becoming increasingly difficult because peers are becoming increasingly scarce until the day when there will be nothing left to minering.

Bitcoin is a type of database that solves the multi-master replication problem of distributed systems because it has mechanisms that prevent conflicts, much the same as blockchain technology that ensures the spending of single outputs more than once.

Well, the transactions with Bitcoin allow a publicly auditable per-row permission scheme. Therefore, the problems that Bitcoin is experiencing are challenges for science and technology that are looking for ways to solve.

The reason is that in addition to security, reliability, and freedom it has already been verified that this technology means the materialization of human rights by providing greater dignity since it democratizes the wealth for all people and especially to the isolated people or people where the banks dominate the economy as in developing countries.

The right to Internet connection to communicate or access means of work provides human dignity since it allows the individual to belong to the global community or local fully. This also gives respect and recognition to the environment that each individual has chosen to live and develop himself/herself.

The cryptomoedas embody the real potential to generate desired consequences in the world market by reducing poverty and making traditional banks change their behavior relative to governments and people because they are experiencing competition on this scenery.

We are experiencing the tyranny by banks; undoubtedly, this is incompatible with democratic ideals. With competition, traditional banks will take their place: an agent of financial services in the economic sector and nothing more. The individual gains the freedom to manage his/her own financial life and frees himself/herself from oppression.

The raison d'être of Bitcoin is to prevent the monetary tyranny [7] at the moment the society was approaching a breaking point. Essentially, Bitcoin is a reaction to three separate and ongoing developments: (a) centralized monetary authority, (b) decrease of the financial privacy, and (c) entrenched legacy financial infrastructure. Bitcoin emerged as a natural response to the collapse of the current monetary order, to the constant reduction of financial privacy, and to a banking architecture increasingly damaging the average citizen. Governments cannot inflate bitcoins. Governments cannot take ownership of the Bitcoin network. Governments cannot corrupt or devalue Bitcoins either.

It adds to the above idea the analysis made by the World Bank Financial and Private Sector Development Consultative Group [8]; their financial experts believe that the blockchain's technology and crypto-coins are a potential weapon against poverty and oppression. Bitcoin also has the potential to improve the quality of life of the poorest people in the world. Increasing access to basic financial services is a promising anti-poverty technique. According to estimates, 64% of people living in developing countries have little access to these services, perhaps because it is very expensive for traditional financial institutions to serve poor and rural areas.

The Internet is the world's largest computing network. There are two important namespaces that allow us to find resources on the Web: the IP address system and the domain name hierarchy. The DNS is the global infrastructure that maintains the domain name hierarchy coherent and manages the translation of domain names in the related IP addresses.

The registrars are authorized entitled to register domain names in a particular TLD to end users. When the registrar receives a user's registration request, it verifies if that name is available by checking with the appropriate registry that manages the corresponding TLD. If it is, the registrar proceeds registering the name with the registry. That, for its part, adds the new name to its registry database and publishes it in the DNS.

As CASALICCHIO [9] is saying us, were in the technological age that it will increasingly be based on ICT networks. In this field, it is clear that DNS plays more and more a strategic function in maintaining reachability of all nodes of this large, distributed system. Then, there is trully possible a Blockchain distributed ledger from recording the same names, information and IP Addresses for all internet based services that DNS records today. It means more accountability coming up as a helpful tool against crimes into system such as laundering money.

When we are thinking about if it will be possible without a governing body like ICANN, we will conclude that anyone could propose a name with or without a TLD equivalent, and associate an IP address with it. No two people would be able to own the same domain, and there is nothing to stop domain names being traded as they are today.

But intelligent solutions make innovative use of ICTs to improve quality of life, efficiency, productivity, and competitiveness of services. We can see more and more about IoT interconnecting devices, objects, and people together with AI to analyze and execute the huge volumes of data that follow. Thus, moving DNS to a distributed blockchain platform is technically possible.

At present, developed countries are already working to establish international standards for interconnection and communication of devices and establishment of interoperable, reliable, and transparent services on the global scale.

This means relevant points when we see countries like China investing in smart economy and especially in the Eurasian Corridor, the East European continental extension to China, where blockchain is the vehicle developing the smart economy in that region. This has been interesting to the oil-producing countries negotiating through smart contract run away from economic sanctions in dollar deals.

A report released by the Cyberspace Administration of CHINA [10] said that the size of China's digital economy grew to 27.2 trillion yuan last year, up 20.3% year on year and accounting for 32.9% of the country's gross domestic product.

Thus, IP-based networks are a key component in the Internet of Things (IoT) as it is the technology to offer cost-effective connectivity.

Whereas the 2030 Agenda for Sustainable Development from the Department of Economic and Social Affairs by United Nations is predicting that increases about devices connected to the Internet will treble in next 15 years, DNS servers as a means of strategic point to financial lives of people because it is responsible for locating and translating the addresses of websites we enter into web browsers into IP addresses. Smart city ICT infrastructure needs to support openness and interoperability, which will only be achieved with coordinated adherence to international standards.

IPv6-based networks bring the virtually unlimited addressing scalability required for smart cities.

In the past 20 years the global economy has changed rapidly. Particularly marked has been the development of world economic integration and standardized products. I add to this scenario the following: (a) the capital movement was more prioritized than the trade coming up as the driving force of the global economy; (b) production was prioritized than employment; and (c) primary products have become prioritized than the industrial economy.

This new economy has brought direct consequences to dynamics of capitalism, doing many companies to carry out the commercial activity in the virtual environment by offering INTANGIBLE GOODS AND SERVICES. In the other hands, the taxation models didn't keep the same pace than transformation promoted by Fourth Industrial Revolution. One of the most difficult problems has been to define the tax residence to a company, or even the identification which incident tax under a fact of economic relevance. To define the Permanent Establishment is a difficult knot to untie.

Despite that, the benefits acquired by technology is improving some sectors such as agriculture to perform higher productivity through mechanisation but this new scenario brought the decoupling of the primary product market from the industrial economy and the consequences are: less jobs and less money to mostly people and underdeveloped countries. It happens because this countries finds obstacles to deploy the technological education in short time. The international cooperation is a fast way through the development of partnerships to help developing countries walk in the same pace than the technology. Through partnerships, education and international cooperation more jobs and better education will arrive to everyone and the Global Economy will accomplish the goals to put end in hunger, poverty and inequalities.

From my point of view I argument that the domestic economy migrated to the world economy as the chief economic unit where Europe and Japan are prominence countries and the developing countries are at the mercy of world supply and demand movements, with the resultant fluctuations in prices.

The nature of economic transaction at all levels was changed when digital economy came up in the world economy because before that the nature was money for money in unequal and discriminatory biz. By now, there is a social value in economic transactions able to touch all social levels. The digital economy with blockchain is completely nondiscriminatory and a revolutionary phenomenon.

4   The Method to Solving the Tax Complexity for Smart Economy with Blockchain and Cryptocurrency: Building the Regulatory Framework

This research is innovative because it creates the 6th dimension of human rights as a scientific discovery. This means one more layer of fundamental rights for the survival of the human being which is added to the set of rights protected in the Universal Declaration of Human Rights, 1948 [11], which guides the common norm to be achieved by all peoples and nations.

Human rights are also guidelines that impact how the law will be treated by the Organisation for Economic Co-operation and Development (OECD) and for these reasons the research is useful and innovative to mankind [12].

The Organisation for Economic Co-operation and Development (OECD) is an intergovernmental economic organization with 36 member countries, founded in 1961 to stimulate economic progress and world trade. It is a forum of countries describing themselves as committed to democracy and the market economy, providing a platform to compare policy experiences, seeking answers to common problems, identifying good practices, and coordinating domestic and international policies of its members.

The scientific discovery adds to the knowledge as it establishes a legal and socioeconomic doctrine that guides all practitioners working in the digital economy by being government, legal, or private sector members.

It represents an important discovery because fundamental rights are the guidelines to the drafting of laws. And specifically to the object proposed in this research, establishing technology as a fundamental right, treated exclusively in the sixth dimension, will serve to elaborate laws that aim to maximize the application of the rates. To the authors, high taxation, laundering money, and frauds are leading humanity to widespread impoverishment.

The humanity will get benefits from this legal thesis because it will be possible to apply lower and fairer taxes to digital economy and this way the world economic flow will be increased faster. For smart economy it will be possible to establish a tax exception in compliance with law, norms, and legal and humanitarian principles.

The publication of the research will allow the discussion into the technical and scientific community and will provide an overview to the inclusive standard in the regulatory framework in a humanitarian approach.

The peace on physical borders will be a reality if governments assume the joint reliability for human security by making the joint effort to establish taxation in the smart economy with a humanitarian focus.

The stability of the humanity has layers that overlap: food, health, legal, social, and healthy environment. And the economy is the armor that holds stability because it protects and provides means of concreteness of the layers of stability. Politics matters only insofar as it serves to structure and maintain stability. The borders are protected and global peace is established when people don't need to pursue survival in other countries.

The solution lies in education and partnerships between developed and developing countries to accelerate the economy through emerging technologies that are bringing more jobs, and new ways to develop business and financial life autonomy. The partnership is the key to accelerate all because it requires low initial investments to establish a profitable company.

There is a need for countries with established smart economy to help others. This will keep the physical borders protected because since 2008, with the establishment of the smart economy, borders are not physical, and globalization has established the virtual borders.

Countries with a strong economy will not invade others in search of survival. The biggest example about it occurred in 2018, when 2000 Mexican children ended up caged in the United States when their parents tried to illegally enter there. The sovereignty of the American Government had laws that allowed this, but this characterizes crime against humanity (source: Carta Maior newspaper) [13].

Another equally common situation in Europe is Libyans entering the Mediterranean trying to reach countries like Italy, Portugal, Spain, and others. Most of them die on the crossing of hunger and thirst. Others get to the destination and face a life of legal immigrants. But it is the price of survival (source: Época newspaper [14]).

There is no power without money. Even the toughest rulers and dictators have realized this and invest in diplomatic, political, administrative, and social infrastructure because war creates weak economy and power interests only comes correlated to economic power.

For these reasons there is already a joint desire to invest in disruptive technologies to strengthen the economy around the world and this legal doctrine is bringing an innovative contribution built through the results obtained with the research.

In turn, it will become possible to understand the interaction of norms, laws, and taxation on the cryptocurrencies as an innovative socioeconomic system.

In line, there is a need to build the meaning and content of the transactions with these currencies to be useful like an argument in petitions in the international courts. This will be done either now creating or sometimes innovating or sometimes extinguishing consolidated legal understandings but which for us will not be useful because the legal requirements to digital economy start from a unique approach if we are comparing with the classic models applied to financial systems or business.

In this way, through the constructivist process it will be possible to direct the governance relationship with such coins without detracting from the purpose for which they are intended.

In order to achieve the purpose of blockchain's creation and payment transactions with cryptocurrencies, the government authorities should concentrate their efforts in building a minimum rule to guarantee security, a certain privacy, and maintenance of the system.

Recalling that the practical result got from all tools posed for monetary policy purposes is basically the manipulation of the money supply in the economy. To foster a stable and sustainable economic activity, banks manipulate the compulsory level and open market operations and the various regulations issued by the monetary authority of each country. Currently, the manipulation of the money supply occurs indirectly, by the direct influence on the interest rate.

This economic strategy benefits governments and banks unlimitedly. So, when people claim about the necessity about the regulatory framework to smart economy, specially for the cryptocurrencies, the rulers in developing countries are ignoring because it means a political decision that doesn't matter to these economic and social agents. The emerging economy threatens their power.

On the other hand, they would be interested as long as it will be a marker between victory and defeat in the elections. The popular pressure passes through the democratic ideal of freedom, and for such the system of cryptocurrency is there and each individual can choose to use or not. So, electors will support the candidate who would agree with the system.

However, the political interest gives signs of life on the part of the rulers who are commanded by the banks and they are aware of the popular pressure and the potential for taxation. The popular pressure is defined by the impotence of the governments in extinguishing this social event of economic reflexes that has its own independent life of state performance.

A minimal rule would only come to make transactions safer. It is not a matter of obedience to spurious interests threating the public good. If government authorities agree with the financial system of payments within blockchain through virtual currencies, they will know that state powers to manipulate the economy and oppress the people would be weakened. The proper rule finds obstacles and walks through crooked alleyways. Understanding economic policy events is not something simple for the average citizen. The ideal of financial freedom to be achieved by the political path is conditioned to the education of society and represents a critical point in the democratic process.

The necessary notes were made and it allow us to start making the trajectory to build the main line to matrix rule of tax incidence respecting the social event, function, and appropriate content.

From now, the signs that are performed in digital economy will get the important role to establish the regulatory framework in an inclusive standard. For example, rulers will interpret the signs at the moment that they are making resolutions, protocols, or laws. Signs applied to draw the matrix rule of tax incidence means everything that is causing impact like cloud environment and hardware means the spatial requirement as well the miner or investor who is liable for tax.

Taking this text as our material essence, this composes fit physical support that will lead us to the meaning that the presented event forces us to interpret. Thus, to fulfill such desideratum, the signs are semantically related to the objects meant. These form the necessary framework so that we can contribute to the theme by decoding it until it reaches useful rule to enable building the matrix rule of tax incidence (MRTI).

In other words, this will be the trajectory that generates useful meaning in the international legal community for taxes and fees in a pragmatic view, with a small portion of logic according to the competent language that conforms to the international laws and those intrinsic to Brazil in order to allow the appropriate legal subsumption.

The social event introduced a relevant fact to the science of law, which when dissected proved to be interesting to the legal community and, especially, to the tax law. From the system referred to the cryptocurrencies, we take important signs similar to the traditional financial model to draw up proposals and extract the necessary content to fill stricto sensu legal norms with legal meanings in the hypothetical-conditional form in which the normative content can be systematized.

It is intended to outline the elements needed to build the matrix rule of tax incidence (MRTI) for financial transactions within blockchain's technology with the virtual currencies. In order to avoid semantic inconsistencies I note the dichotomy between "broad sense" and "strict sense" of the norm.

The former denotes units of the positive law system, even though they don't express a complete deontic message. The second denotes the complete deontic message, that is, they are meanings constructed from the statements put forward by the legislator, structured in the hypothetical-conditional form [15].

This is the mandatory trajectory to support this economic sector brought by blockchain technology and cryptocurrencies (the first one cryptocurrency), preserving its essence and function. If we steal from this path, we would fall in a improper method just seeking simply mutilation of some of MRTI criteria for forcing the "fit."

The MRTI is characterized as a logical method of organizing the text of positive law. It is the formula that organizes prescriptive enunciates extracted from the legal texts for later interpretation and evaluation of the constitutionality of the taxes. In fact, the MRTI is a rule of pre-ordered behavior to discipline the conduct of the taxpayer relative to the subject holding the right to credit: government authorities authorized to receive the tributary obligation.

The schematic and operational mode of MRTI results from the unfolding from "logical-semantic constructivism," leading to join the theory and practice of the investigation of the tax object to unravel the legal tax. In short, it is a legal rule in the strict sense, because its core is essentially the definition of the general and abstract rule able to identify the minimum elements of the tax rule. It means a hypothetical-conditional judgment that conveys a hypothesis and a consequence in details.

Therefore, the entrance of blockchain technology and the cryptocurrencies is an event occurring in the world requesting a description in a competent legal language to establish the tax obligation link and identify the taxpayer as a taxable person and the state as an active subject.

So, does the state have any right to demand some benefit from this economic chain? Would the participants in this economic chain be taxpayers with the duty to provision the tax liability? If cryptocurrencies are essential for freedom without interference from the state or banks, is there a link to establishing the tax liability?

In the sequence the answers will come up because the assumptions adopted will allow the filling the necessary content for the answers. This approach considers the tax incidence seeing to focus on subsumption and implication. Now, our first problem concerns about the legal behavior when the taxation in some countries is considering the transactions within blockchain only as financial assets.

Logically, in establishing subsumption, the event examined is recognized as an integral element of the class of facts predicted in the supposed general and abstract norm. However, when treating the transactions with Bitcoins with the blockchain technology only as a financial asset subject to income tax, the MRTI is unequal from the general and abstract norm because the material criterion isn't the same.

The reason why there is the imperfection in normative framework when classifying just as a financial asset is that current laws and regulations have not yet dealt with all applications that are happening inside digital economy; innovative technologies such as artificial intelligence and smart contracts brings to scenery some legal gray areas.

Firstly, this is because Bitcoin, Ethereum, or other cryptocurrencies don't fit into existing regulatory definitions of currency or other financial instruments or institutions, making it complex to know which laws apply to it and in what form. And secondly, we are faced with complex operations with varying results so a single or excluding view is not correct.

Bitcoin and all other cryptocurrencies are of legal nature as sui generis because they have properties similar to electronic payments, but sometimes they conform to commodities and other currencies. The degree of complexity in this performance is high.

The regulatory framework has become one of the biggest debates in all countries because the smart economy is growing rapid and solid as an alternative to classic financial system.

In the United States more than 60% of American companies are establishing ways to viable operations with the virtual currency that isn't limited to Bitcoin. Japan in April 2017 passed a law making Bitcoin a legal payment instrument.

In turn, in Denmark the government and the financial supervisory authority have announced that business with Bitcoin will be taxed in the same way already as existing taxes, but that taxpayers will not be subject to taxation when performing foreign exchange transactions. Even Denmark's central bank is considering digitizing the country's currency, which would be called e-Krone. In Germany and most of the European Union countries, the German Federal Ministry of Finance has a position that Bitcoin should be seen as a unit of account and private money and therefore should be subject to sales tax and IVA (tax to asset increase).

China is leading in a positive way to establish a health environment to digital economy but specialists are saying that this regulatory framework has a nocive censorship. China's Internet regulator, the Cyberspace Administration of China, CAC [16], made a final draft of regulations concerning cryptocurrency and blockchain companies. The rules will come into effect starting February 19, and provide a set of guidelines that blockchain companies are required to follow.

Regarding this Chinese framework the blockchain will be used to seek for irregular or illegal companies. Blockchain service providers are required to register with authorities within 10 days of providing the service. This process is done through a blockchain information service management system.

Once the filing process is complete, authorities will assess and enter the company into records within 20 days as well as the companies are required to publically display their record number.

In Brazil, public hearings have been held since July 2017 and this resulted in Bill n. 2303/2015 which provides for the inclusion of virtual currencies and air mileage programs in the concept of "payment methods" under the supervision of the Central Bank (amending Law 12,865, of 2013 and Law 9.613, of 1998). The situation of Bill 2303/2015 until April 2018 received the legal advice to constitutionality, legality, and legislative technique but there was no pronouncement as to financial and budgetary adequacy, and, on merit, by approval.

However, there is a concern to note about this; the Bill can become a way to manipulate and to appropriate this technology. This way, it will be possible to subject the control over blockchain and cryptocurrencies under government in the transactions with Bitcoins and by blockchain. This means a point of view from a person that is worried just about getting benefits to himself/herself and not for collective purpose because the strategy allows the control of the Central Bank.

In defense to this point of view, the United Nations and its telecommunications agencies consider that everything is happening among DNS and digital world as part of the telecommunications system and this is one of the attributes that makes the digital economy with blockchain and cryptocurrency or applications about artificial intelligence, big data, IoT, and other technologies an international property protected by human rights as a material instrument of freedom propagated by a change in the content of the communication and therefore impossible to be appropriated by a country.

The Department of Economic and Social Affairs of the United Nations through the Development Policy and Analysis (UN, 2018) division is in the process of adopting blockchain's system and cryptocurrencies through the use of Bitcoin and Ethereum as a way to reduce poverty. There is currently a World Food Program (WFP) among these measures and has been organizing a pilot test, scheduled to begin in Jordan where the WFP will send an indeterminate number of dinars to over 10,000 beneficiaries who need financial support and extra food, with the goal of increasing the number of receivers to 500,000 people still in 2018.

The gray area has been delimited and considerations about international legislation and regulation are noted, as well as the status in Brazil for blockchain technology and its role in the cryptocurrency system. At this point we are able to verify the possibility of constructing a tax incidence matrix rule (MRTI) for this complex system.

5   Dialogue Between MRTI and Blockchain Technology

There is one invisible factor in everyone's eyes within the digital economy: the ability to generate profits from communication.

Smart contracts has eliminated and solved many human-caused problems.

Before the technology, the employees and decision makers talked to each other to know about the problems inside the company and get solutions.

When email communication came up, the people didn't talk much anymore.

Nowadays, it is common to see employees and decision makers ignoring emails because they have decided that it wasn't important. But this decision doesn't means any communication model because there was no consensus from the other side.

Thus, companies take losses and even break down because problems are not solved.

At this point, the digital economy and smart contracts represent the solution to a society that has stopped talking.

To continue, we must adopt a useful concept for cryptocurrency that is by definition is a type of digital currency based on cryptography, or the process of converting plaintext into ciphertext, thus making readable text non-decipherable. The use of cryptography in the transfer of data has four main objectives: (1) confidentiality: the information cannot be understood by anyone for whom it was unintended to be; (2) integrity: ensuring the information sent remains unaltered; (3) non-repudiation: the sender of the information cannot deny that they sent the information at a later date and time; (4) authentication: the sender and receiver have the ability to confirm each other's identity and the origin and destination of the information [17].

Currently in Brazil, the Federal Revenue Service has treated virtual currency as a financial asset. There is no adequate legal framework considering the different results from transactions and, consequently, there is no legal rule foreseen for the conversion of the amounts for tax purposes. It is recommended by the Federal Revenue Service that the financial asset must be declared, and will suffer incidence of the income tax (IR) by disposal of assets, because this is the moment of the equity increase, considering the value of the acquisition. By equivalence, it will follow all or relative rules about income tax to companies or personal, according to the case. ITCMD (tax on the transmission of assets by death or donation) will be possible and it will also follow rules when donations are possible to be proven by skillful documentation.

But this tax treatment isn't correct because there is no conversion from crypto-currencies to real (BRL). And applying tax rates and laws typical to the classical economy for the digital economy violates all legal and humanitarian principles. Moreover, it violates the requirements of MRTI because the commercial and financial transactions that occur with the digital economy have different motives and sources.

Brazil's justification for acting this way is that there is no possibility of the Bitcoin wallet being tracked by any country and therefore the Federal Revenue Service hasn't got resources to verify the quotation or historical values. Only the equity increase that goes to the classical financial institutions will be visible.

Brazil is used as an example in this text just because it is a country of great territorial and populous extent where poverty is making victims every day. But there are other countries doing the same thing. It doesn't sound like reasonable that the large, populous, and poor countries should not be interested in developing social and economic policies that have demonstrated a great capacity to reverse the negative background.

It is noteworthy that there is already a virtual and international business field where Bitcoins can be consumed in hotels, parks, and restaurants where the exchange takes place in the blockchain market without going through any financial institution. In this case, the equity increase is invisible, but not illegal. There is no possibility of criminalizing an event that isn't described in the law as a crime, Section 1 of the Brazilian Criminal Code.

As an example about the business with the blockchain and Bitcoin technology, there are hotels being built entirely with resources from ICO and in countries where there is no influence of antitrust laws. Such a procedure isn't typical of money laundering laws because there is no illicit activity.

In the exemplary case above, the building of a hotel with financial resources was acquired through investments in ICO, and then it will be operated with tourists who hosted in this hotel and they will pay for their daily expense with the wallet of Bitcoins. Well, there are the following transactions: credit and service operations through the exchange and not the typical sale as it is currently configured.

All financial transactions occurred virtually with the exchange within the blockchain database where computers make various trading operations solving mathematical problems to find the correct algorithm that results in Bitcoin rate. In this case, there was no physical asset increase to be checked and there is no legal provision to investigate crime for lack of illegality.

Obviously, rulers have already thought of legislating to criminalize such transactions, but this would distort the system and would be legislative efforts depleted by the death of blockchain's own financial system. It concludes the idea that they want the death of the system. Initially, I said that market forces leave governments and banks with their hands tied and powerless. Market forces are obstacles that prevent the death of blockchain technology and cryptocurrency.

The system is independent, freely created without increasing costs or damaging the governments. Included, it received support by the United Nations as a way of eradicating poverty. According to Andrea Garcia, 2016, the financial system provided by blockchain technology means a democratic concept about the "Tape to Contention to State Greed and obliges them to fulfill the Public Commitment."

The system is only subject to ICANN intervention because the root's zone addresses moved by DNS registers are the basis for blockchain's move. However, this is minimal control since ICANN doesn't address the content of messages.

From the researcher' point of view, in relation to any digital resource that is labeled as cryptocurrency, to establish the tax legal nexus able to build the Matrix Rule of Tax Incidence to each transactions, it will depend on the characteristics and use of that resource analysed one by one in particular. Since there are different ways to investments through digitalised economy with different cryptocurrencies and tokens to specific purposes, the regulatory framework should treat differently each situation and establish the Matrix Rule of Tax Incidence considering differents criterias.

Carefully considered, about the scenery inside transactions with blockchain and Bitcoins, there are requests from individuals who operate or mine in the ICO market desiring the minimum rule to comply with the rules of compliance depending on governance activity. The reason is that transactions have principles that are being followed by those who operate responsibly: freedom, security, soundness, reliability, accountability, and integrity.

July 4, 2018, marked a regulatory framework to Maltese Parliament that has officially passed three bills into law, establishing the first regulatory framework for blockchain, cryptocurrency, and DLT (distributed ledger technology): (a)Virtual Financial Assets Bill; (b) Malta Digital Innovation Authority; and (c) Innovative Technology Arrangements and Services (ITASA BILL). Malta was the first country in the world to provide an official set of regulations for operators in the blockchain, cryptocurrency, and DLT space.

This concept of cryptocurrency conforms perfectly to the content and purpose of the system of blockchain and Bitcoin like other cryptomoedas also and without mutilating the functionality. These are exchange transactions that result in the acquisition of Bitcoin or other cryptocurrency and it will only be securities when traded in that market through stock exchanges and described in securities.

The technological innovation brought by blockchain technology to provide financial transactions with virtual currencies, Bitcoins being the best known of them, as an object of analysis focused on the socioeconomic event (disruptive theory and smart economy), describing it in legal language competent to know the legal tributary fact allows the construction of the interesting sense to the law science under the prism of the international tax law.

The MRTI proved possible when we admitted the existence of the mandatory legal nexus between the state and the taxpayers within the blockchain through legal requirement if, and just in case, we are admitting that the governing authorities can tear apart the primary public purpose and they will replace it by spurious purposes threatening the collective good.

The basic premises for building the appropriate Inclusive Standard to digitalised economy through the Semantic Logical Constructivism as a source to defining the tax method, after the methodological cuts necessary, it is stating that when the transaction is a matches to MRTI, it means that the transaction is assuming the characteristics of the securities, it must be registered with the Securities and Exchange Commission respective to each country.

About services, it can fit into the concept of "information exchanges" because the circulation through the Internet is a communication (database) by the technology of blockchain making exchanges all the time and certainly the result will be the execution of an activity, but the MRTI's material core is the exchange and not the obligation to make singly.

However, if there is a mix between virtual currencies and the current currency of a singly country by banking institutions within the transactions, the MRTI's material core should follow the tax laws of physical money because the MRTI's material core loses the intrinsic characteristics of the cryptocurrency when it involves classical banks in bonds, such as independence; transactions without involving banks or state performance; and exchange ratio replaced by financial transaction.

After these considerations, it is possible to create a MRTI for this financial system with blockchain technology for transactions with Bitcoin or other cryptocurrency with criticisms of the legal nexus to establish tax liability.

At the moment, after analyzing the social event and building the sense and content for the legal tax fact, it is possible to fill the topics of MRTI to Bitcoin (or another cryptocurrency) transactions by removing the complexity of blockchain technology.

To the previous concept adopted about cryptocurrency, it will add a limit to tax that was drawn in the American Bill of HR 3708 and considers a value for tax exemption guaranteeing the volatility of the activity and freedom with certain limitations to fulfill the rules of compliance. The ceiling is $600 like a social application.

This is a method that will serve as a basis to build the MRTI specific to each type of transaction or business within the smart economy.

To antecedent in MRTI we would be instituting: material criterion: "exchange" mined algorithms (verb + complement). Considering the verb "to exchange" we are considering singly the computational activity in the search for the correct algorithms. Therefore, there is no increase in equity to justify income tax at that time. Spatial criterion: blockchain. It is a virtual communication system within the Internet network. There is no financial institution involved in justifying the tax on institutional banking transactions (IOF) or a specific country hosting the transaction. Although blockchain uses the Internet, this is the way technology connects to move around. This approach differentiates transactions with Bitcoins within blockchain from other transactions with banks that also use the Internet as a means of accomplishing the task. Temporal criterion: the moment the service fee in Bitcoins arises.

It is the moment of the last signature (hash) in which Bitcoins will enter the wallet for the miner. This will occur because, at the initial step in chain, the miner shares an address indicating the final recipient.

To consequent in MRTI we would be instituting: personal criterion: tax agency (active person and authorized to receive the payment from tax payer) and taxable person. The active person (tax agency) will be the federal revenue body because it is an international transaction where the federal government has exclusive competence to legislate and it may impose tax on international financial circulation. I shall write down my criticisms of the obligatory bond instituted in this way. There is a grave legal harmfulness that concerns to legal nexus to impose taxation. Taxable person can be the miner who acquires the rate and prizes of Bitcoins or other cryptocurrency. Quantitative criterion: calculation base and aliquots. Calculation base will be defined by the Bitcoin rate paid to the miner. Aliquot should be marked by the values related to human rights and the limiting imposed by constitutional and tributary principles able as an obstacle to the state force to tax with excessive greed.

The mandatory nexus able to impose the tax arises with existing laws within the country that justify the collection of abusive taxes to maintain the activities of the state. Remember it: constitutionality and morality may be different from legality when laws are instituted in the light of parliamentary corruption.

One of the juridical natures to justifying the existence of smart economy is to attend to the greatest democratic ideal: the public will.

The blockchain technology is independent of governmental performance, which means that there is no obligatory nexus required to tax when the cryptocurrencies don't enter the traditional financial system or add costs to state possible to place harmfulness to public interests. Then, it is necessary that the government give up of taxation's hunger to make the economy stronger. This way, the country will increase the gross national product (GDP) through the stronger consumers paying taxes when they are paying for services and products.

The legal nexus just will be possible when there is the mix between cryptocurrency and financial institution and bonds. This is because the legal nature of the transactions within blockchain with Bitcoins is humanitarian and this will avoid states and banks from enriching themselves at the expense of humanity's poverty. The investments made by speculative companies must be treated differently from the citizen who invests in the long term with the objective of self-management of the financial life in an independent way because it will increase the GDP.

Initially, it was stated that the digital economy with blockchain and cryptocurrency, artificial intelligence, IoT, big data, and all emerging and disruptive technologies means a much stronger conceptual idea than the idea that understands them only in the technological field. It was also stated that for this reason it deserves the protection of the 6th dimension of human rights. The researcher also brought solid foundations to describe the tax regulatory landscape as a decisive marker for technological evolution. The next lines of the text are dedicated to justifying and bringing evidences that underlie the theory that was proposed.

6   Dimensions of Fundamental Rights: Universal Right to Technology as the 6th Law of Protection

There is a new social contract increasing digitization, new platforms, digital business models, and non-state actors exerting considerable pressure on government and its legitimacy, creating implications for public policy on issues ranging from taxes to basic accountability. Advances in technology and continual re-tooling create the need for new forms of social protections but also create the potential for more efficient service delivery. From researcher's point of view, increasing urbanization will accelerate innovation but also creates economic shocks.

The economic shocks can be solved with investments in framework regulatory and establishing broad and universal access to technologies like fundamental rights.

The relationship between the state and individual has long been given definitions and classifications both to explain and to establish limits about the interference from the public interest into the private interest.

Human rights are inherent rights to all human beings, regardless of race, gender, nationality, ethnicity, language, religion, or any other condition and it includes the right to life; the right to liberty: opinion and expression freedom; the right to work; and the right to education and many others. All people deserve these rights without discrimination.

International human rights establish the obligation to governments to act in determined ways or to abstain determined acts in order to promote and protect the human rights and freedoms to groups or individuals.

Jellinek's theory is concerned with classifying the limitations to which the conduct expected by individuals and rulers is subject.

Created at the end of the 19th century by Georg Jellinek, an important jurist and philosopher, Jellinek's Four Status Theory indicates four conducts that an individual could show to state: passive, active, negative, and positive.

Passive status (status subjectionis): the individual is subordinated to the public powers, being the holder of duties in relation to the state that imposes its will through laws that allow or prohibit conduct.

Active status (status actives civitatis): the citizen exercises their political rights. Thus, there is the possibility of the individual interfering in the will of the state, for which among other ways we have the vote as part of this.

Bobbio [18], Italian political historian, in 1998 already addressed the decentralization of state power over the individual as healthy under socioeconomic aspect and as a way of making democracy sustainable.

Negative status (status libertatis): indicates the freedom of the individual in relation to the state, he/she will act if he/she wants, in some situations, released from the performance of the public power in an exceptional way. Example: freedom of expression.

Positive status (status civitatis): possibility of the individual to demand of the state about some provision, and the public power must act in a positive way in favor of this individual. Example: the possibility of the individual requiring the right to health.

From this explanation, the importance of the methodological cut is explained by the fact that historically human evolution brings new facts for legal science to harmonize the legal structure with human needs.

Often the collision of rights between free initiative and the autonomy of the private will and the dignity of the human calls for the maximum effectiveness of fundamental rights as a sustainable way of guaranteeing future generations. In case there is fundamental rights collision, it is recommended to use the judgment about interests in a reasonable and harmonious way within the system of laws of each country, or in an international environment when applicable.

In the study about human evolution, the fundamental rights are given layers of protection that interact harmoniously with rights and duties individually as well as collectively.

Legal doctrine usually divides them into generations or dimensions. The authors prefer to adopt the term dimension because it better reflects the legal treatment that overlaps history in the form of layers. However the term generation passes the idea of a right that gives way to another in a successive way. Therefore, establishing technology as a fundamental right is important as it ensures constant human evolution in all its aspects through the application of AI (artificial intelligence), ICTs, IoT, and others that can be combined with blockchain technology for smart economy with cryptocurrencies.

We have five (5) dimensions of rights enshrined in current legal doctrine [19].

The first dimension is a landmark about the shift from the authoritarian state to the rule of law. They are rights related to public liberties and political rights, that is, civil and political rights translating the value of freedom. The historical documents are (a) Magna Carta 1215, signed by the king João Sem Terra; (b) Peace of Westphalia (1648); (c) Habeas Corpus Act (1679); and (d) Bill of Rights (1688) and the Declarations of American (1776) and French (1789).

The second dimension was driven by the European Industrial Revolution of the nineteenth century where the poor working conditions represented the motivation for civil revolutions. At the beginning of the twentieth century, World War I showed the need to establish social rights. The second dimensions are social, cultural, and economic rights corresponding to the rights of equality. The historical documents are Constitution of Weimar (1919) in Germany and Treaty of Versailles (1919) at the OIT.

Third dimensions are transindividual rights that have resulted from the shifts that the international community and the mass society have undergone. It is the phenomenon of globalization developing technology and legal science to establish the right of communication and the property right over the common patrimony of humanity. But economic and sustainable development wasn't the goal. The focus was on the rights of solidarity as the right of everyone to live in a healthy environment.

The advance of industry is responsible for several companies emerging and thus the need also arose to assure the consumer protection mechanisms making the state responsible for the establishment of norms to regulate consumer relations. The state has the duty to guarantee consumer protection through public policies and access to specific bodies that can solve the problems arising from the commercial relationship.

Documents: (a) Consumer Bill of Rights (1962); (b) Council of Europe (1973); (c) Commission on Human Rights (UN), in session 29 recognized the fundamental and universal rights of the consumer; and (d) Resolution 39/248 (1985), UN.

The fourth dimension means individual rights introduced by political globalization. Highlights include the right to direct democracy, information, and pluralism. Fundamental rights belong to an autonomous dimension and that must be globalized in the institutional field.

Fifth dimension is about the right to peace: it is from the right of collectivity to guarantee the human rights and the peace is the axiom of participatory democracy, with right to life being an indispensable condition for the progress of all nations and in all spheres. Fifth dimension guarantees all peoples the right to self-determination. It is the right of a people not to submit to the sovereignty of another people (or governments) against their will. It is the possibility of this people to determine their destiny, and their cultural, social, and economic development, without the influence of another. However, it finds the limit that in the right about, in no one case, a people will be deprived of its own means of subsistence.

The author points out that currently most countries are committing to human subsistence when taxes mean about 50–80% of taxes payable to the state [20]. It is a confiscatory encroachment when governments hold such a large share. This is one of the causes that is raising poverty rates and starving people. In this step, the authors warn that the world will collapse.

Documents: Universal Declaration of Human Rights (1948, UN) and Declaration on the Right of Peoples to Peace, contained in UN Resolution 39, 1984.

Sixth dimension of fundamental rights [21] was created in 2015 by the author, ROMAOLI—lawyer and legal scientist—and presented for the first time in 2018 at the International Conference on Cognitive Computing WEBbr 2018, held in Brazil:

The sixth dimension arises from the 4th Industrial Revolution and incorporates the right to social and technological development as a fundamental right that allows the individual to self-management, self-development, and self-concept. It concerns the historical period started in 2009, when the first virtual currency (Bitcoin) was introduced to the economy. It brought to the citizen greater financial independence joined to the payment system introduced by blockchain technology in ecosystem without intermediaries such as banks. Sixth dimension brings to this century a minimal acceptance of the interference from state in the intimate and financial life of the citizen. Social awareness strengthens the bonds of commercial partnership based on the will to share the gains more quickly and justly than the state is doing it. There is a high degree of collective consciousness where the sense about land borders has been replaced by virtual boundaries. It sounds like a maximized bloc extracted from all dimensions of human rights embodying the sixth dimension.

The fundamental protection is justified on the grounds that the social and economic benefits that technology has provided to all peoples are undeniable because it has a great accelerating power of development in all spheres of society life.

Reliable vehicles and communication officers each is providing reports and scientific research that restate this. It is the smart economy scenario that reveals reasons for concerns, but the advantages have overcome the problems.

One of the evidences about the technology as a fundamental value comes from Kofi Annan [22], former Secretary-General of the United Nations, passed away on 18 August. In the speech delivered at the Massachusetts Institute of Technology (MIT) in 2015, he was sharing his views on how technological advancements could improve the state of the world. He established the technology and political will such as the golden key to solve any crisis.

Renowned researchers point out that technology will take many forms and influence many sectors. Education is an important part of the technological challenge [23].

Along the same lines, the World Bank Group report, 2017 [24], brings the same evidence that it allows to elevate technology as the sixth dimension of human rights to claim universal access to the connection in Internet and technologies when their analysts established the technologies as the driving force behind economic growth, citizen engagement, and job creation. Information and communication technologies (ICTs), in particular, are reshaping many aspects of the world's economies, governments, and societies.

The technology has been recognized by the United Nations as a great transformative potential of society to ensure a dignified life in the 2030 Agenda for Sustainable Development [25] from the Department of Economic and Social Affairs. This Agenda is a Plan of Action for people, planet, and prosperity. It also seeks to strengthen universal peace in greater freedom and recognizes that eradicating poverty in all ways and its dimensions, including extreme poverty, is the greatest global challenge and an indispensable prerequisite for sustainable development.

The Agenda guides all countries and stakeholders to work in a collaborative partnership. They will implement this plan in a joint effort to release the human race from the tyranny of poverty, to heal and protect the planet directed to change the world into a sustainable and resilient path. To achieve this goal 17 sustainable development objectives were established demonstrating the scale and ambition of this new universal agenda. They seek to realize human rights for all and achieve gender equality and empowerment of all women and girls. The goals are integrated and indivisible and balance the three dimensions of sustainable development: economic, social, and environmental.

The goals will stimulate action over the next 15 years in areas of critical importance to humanity and the planet. Objective number 17 is about development of partnerships because it has the capacity to arise the world and the life of each person to a better level. Development of education is addressed in objective number 4 and it is about actions to stimulate and provide for technological education (4b). From these objectives it will be possible to impulse the economy and to perform the objectives from 1 to 5: exterminating hunger and poverty and creating a sustainable environment.

After all the above arguments, which have not been exhausted, is possible affirm the TECHNOLOGY AND SMART ECONOMY WITH AI, BLOCKCHAIN AND

CRYPTOCURRENCY came up as the 6TH DIMENSION HUMAN RIGHTS. This legal thesis was showed at first time at the 2.018 WEB Conference for Cognitive Computing as a full legal doctrine. As a positive result from this scientific presentation came with the approval from the scientific community about the signs that the Fourth Industrial Revolution brought to society and it has been modifying the way citizens are performing the legitimacy into Neo-Constructivist Democracy trying reach out the human security. Thus, the minimum existential is becoming available and the dignify life is a reality from technologies increasing the flow of economic and social development.

Moreover, it promotes a dignified life and creates new professional positions with a high degree of individual autonomy to generate personal satisfaction in the individual, and consequently, throughout society.

The utility to arise the technology as 6th dimension human rights is to solve the problem that could come up if developing countries and poor peoples are placed out of access: the poverty and hunger would keep growing. But like fundamental rights it gains the same status as universal right like health, food, education, and others. This way is the responsibility of governments investing in actions to making the technology available to all.

7   Democracy and ICTS: AI, IoT, Blockchain in the Smart Economy to a Sustainable World

Nowadays the democratic debate confronts challenges within the worldwide social and economic scenario, because the corruption is just one of the serious problems. Without adequate governmental investment on social work there will not be worthy distribution of the social justice. In all countries the social work from government is insufficient because tax money is being spent to maintain the high standard of public administrator's lives or diverted to keep the corruption system. Therefore, there is accentuated regression of fundamental and individual rights besides the stimulus to the civil society's depoliticization process, specially in developing countries.

This economic vision came up to 2030 agenda for sustainable developing that believes it is possible to solve through partnerships between civil society, private companies, associations, and foundations. This partnership model is a showcase about social benefits sharing with the humanity when people get large financial autonomy. The partnership is more efficient than governments because it is more fast and it acts in focal situations without intermediates with less bureaucracy. More autonomy is bigger democratic exercise and it can improve the education through sharing about experiences among developed countries and developing countries besides financial mutual helps.

It is the opinion from other political scientists too. NGOs' role and influence have exploded in the last half decade and their financial resources and their expertise sometimes exceed those of smaller governments and of international organizations. Their influence is irrefutable as we have seen its acts in many countries as they are delivering the service in urban and rural community development, education, and health care—that faltering governments can no longer manage [26].

With accuracy, democracy is an essential principle. Therefore, the social work has an important function of social regulation, as it is a trustee of values like dignity of the human being, self-determination, and social justice and, by nature, represents the ideals of democracy. This allows for peace and lessens interest in physical boundary disputes.

The work developed here focuses on the human needs, thus requiring them to be met, not by a matter of option, but as an imperative of basic justice. Therefore, the social work moves toward considering human rights as one of the organizational principles of its professional practice and which must be preserved by any democratic country that has signed an international treaty aiming at ensuring human rights.

In summary on the guiding principles of democratic states, the supremacy of the popular will underscores popular participation in government decisions like a necessity for holding peace and improving the economic flow to compose gross domestic product.

Another important guide driving the democracy is the principle about the preservation of freedom, understood as the power to do everything that isn't harmful to another person and also it means individual freedom to dispose of himself/herself and property, without any interference from the state. The General Data Protection Regulation (GDPR) is an example of allowing people to sell their own personal information.

Both principles are combined with the equality of rights, understood as the prohibition of distinctions in the enjoyment of rights, mainly for economic reasons or of discrimination between social classes.

There is also a paradox between laws and guarantee of a dignified citizenship compromised by value about popular needs versus possibility of the government providing social programs when 83 million people are being added to the world's population every year [27].

More popular financial autonomy isn't a whim. Indeed it is a necessity to ensure human survival that will be achieved with investments in technologies for broad access.

It will not be possible if taxes are not re-treated by governments that must make a regulatory framework with taxes that are more progressive and proportionate than we currently have. This is about the partnership between governments and citizens driving the administration of the state as a democratic ideal.

As for the existence of a sustainable planet and a dignified life, rulers cannot argue their defense founded on the principle of the "possible reserve" that establishes a limit for the realization of social rights when governments prove insufficient financial and structural resources. The principle is opposed to the existential minimum required by humanity. The existential minimum principle includes everything that is necessary for a dignified life: health, housing, education, leisure, etc. This is why the state cannot deny smart economy with blockchain and cryptocurrencies because the financial interest of the governments is a secondary interest. If the governments are worried about profits, it isn't a public interest. Thus, the state cannot ignore the realization of fundamental and social rights that guarantee the existential minimum for all only because they are afraid of sharing power and the public administration.

The truth is that this is a complex issue. But the complexity can be mitigated through information technology (ICTs) and artificial intelligence (AI) because this technology is maximizing how the human being solves difficult issues. Human intelligence is the ability to know and learn about problem solving. Artificial intelligence is the simulation of human intelligence by machines and includes the ability to solve problems, ability to act as humans, and ability to rationalize behaviors and activities. Creativity is a fundamental feature of human intelligence, and an inescapable challenge for AI. AI models intended (or considered) as part of cognitive science is becoming more and more useful to health like psychology and psychiatric diagnosis. The creativity is a feature of human intelligence in general. It is grounded in everyday capacities such as the association of ideas, reminding, perception, analogical thinking, searching a structured problem space, and reflective self-criticism. It involves not only a cognitive dimension (the generation of new ideas) but also motivation and emotion, and is closely linked to cultural context and personality factors [28].

Current AI models of creativity focus primarily on the cognitive dimension and it is improved by association method and this process will generate the creativity like humans have. But learning machine and external resources like big data when joined to AI perform better than the human creativity.

In modern days, the robots with machine learning reinforcement are now capable of replacing humans at work, assisting in health, and developing business strategies for the financial market but in the higher level than human and of mass-scale performance. In addition, the blockchain technology is showing high performance to hold the integrity in electoral procedures avoiding fake results [29].

Besides that, democracy is being threatened by plutocracy groups and autocratic governments that use fraudulent elections to engage in apparent democratic legitimacy and keep the cycle of corruption and laundering money active.

Meanwhile, hunger has reached millions of people around the world. Only Africa is expected to double its population to 2.4 billion by 2050. On the other hand, Africa has 60% of the world's uncultivated arable land. These are useful information that demonstrate access to technology as a 6th-dimensional human right where there is a need to think about the economy in a decentralized and intelligent way to include developing countries. For example, Africa has the potential to feed the world and to develop through exporting food. They are the warehouse of the world [27].

Two crucial ingredients can help solve these crises: technology and political will. AI, IoT, and blockchain in a society using smart economy will create a sustainable world. But either decision makers must accept the technology or it is useless.

Intelligent digital development means using information and communication technologies (ICTs) to improve people's lives. And everyone knows that beyond positive we have points that present some problems as well. But for the authors, problems shouldn't be an obstacle because it is possible to solve it throughout with one-off measures.

Technology is moving at a fast pace and major technological advances such as IoT, 5G, AI, blockchain, and cloud environment can increase and decrease the gap between developed and developing countries. For this reason, the authors are concerned about establishing grounds for appropriate legislative, economic, and educational treatment.

ICTs are strategically important to smart economy and we definitely need to use them to accelerate the achievement of the sustainable development goals in agriculture, education, and health and also to propitiate more value for societies and communities. Thus, this requires a people-centered approach.

By 2030 we will see an increase in the number of large cities, especially in developing countries, and government programs are required to make possible managing resources to guarantee the minimum existential to citizens. ICTs can help support transport networks and electrical networks, among other key systems for development. If technology exists we must use it to facilitate the future of humanity.

And almost half the world's population is still off-line, and most of them live in remote and isolated rural communities where connectivity is difficult: difficult, for geographical reasons but also because the return on investment in these areas is much poorer than urban areas.

In some countries, the high price of connection service and the tax burden on the service also contribute to this factor. And if we want to keep up with and retain technology benefits, the connection is key.

The researcher brought to the text many reasons demonstrating that smart economy with blockchain technology, AI, and cryptocurrencies are the future of the economy.

By analyzing the scenario of tax law and international policies in the smart economy and also the practical application of technologies such as AI, IoT, IoE, and blockchain working in the cloud environment, it is possible to say that this will lead to real benefits. Virtual is just the environment in which the decentralized economy takes place. The reflexes and benefits are real.

The nature of the economic transaction at all levels was altered when blockchain technology and cryptocurrencies emerged in the world economy. Before that, the financial nature was money for money in unequal and discriminatory businesses and now there is a social value in economic transactions that can touch all social levels. Smart economy with blockchain is completely nondiscriminatory.

8 Tax Morality and Flow of Economic Growth

The next lines in this text report about concerns in the future in smart economy in case if it doesn't receive investments and one-off measures because the dictatorship of banks and the flow of laundering money threaten the world economy to protect the corruption that fuels terrorism and other crimes. The abuse from banks, laundering money, and corruption is feeding growing poverty and hunger in the world.

Dictatorship means all power in the hands of a few: a dictator or a group. This system has cycles because the real power doesn't die out. What can be extinguished is only the power of a dictator or group. Therefore, the power will continue circulating and is disputed all the time.

The society has neglected world history since World War II and today we are living a new cycle of dictatorship that affects the whole world: the dictatorship promoted by the banks through abusing measures against the humanity. Proof of this was the crisis experienced by Europe in 2010, where governments were bankrupt and lived by the orders of the banks. This fact has been widely explored by newspapers worldwide [30].

The ordinary citizens have been borne the brunt of the economy in a lonely way. Note: when banks go into crisis the Rulers give money to keep the activity under the claim of keeping the economy galloping. This is not the real reason because banks don't perform any social function. They don't create jobs. They work for profit by profit. On the opposite, the citizen when loses his job and fails paying the banks, he will be executed and the bank will take his assets and some countries like Brazil it put in risk the survival. In general, governments doesn't give money or employment to make feasible the survivel to citizen in the same proportionality that governments is granting the profits to banks when they is taking losses. This was an example about the cycle of power moving within the economic system of plutocracy.

That is why it is so natural for some countries to resist the digital economy. They are feeling threatened by the individual freedom that smart economy allows by counteracting the tyranny imposed by dictatorship where banks have power and plutocracy is a new economic strategy.

The system of corruption makes public administrators and organizations manipulate the principle of legality to ban most people from all that is good in the world and reserving to plutocrats or some powerful groups: money, food, health, housing, education, and comfort. The authors point out that laws aren't made by minorities or poor people. The laws are made by the powerful and rich people. Anything that threatens the division of power or money will not be accepted by the corrupts. Smart economy gives humanity a chance to get out of this cycle quickly without manipulating laws against the interests of bankers from the classical payment system where they are the middlemen. Currently, banks acquire status as holders of power and controllers of laws. In fact, banks are only part of a country's productive and economic chain and anything beyond this means the abuse against the humanity.

In this regard, smart economy is important because it frees governments and citizens by placing banks in their rightful place: a participant in the financial services chain into country. And only this.

Making fairer laws that limit the power of banks has been impossible to all countries since they have plenty of money to produce laws that favor them. Legality, morality, and constitutionality are different concepts. Not everything that is legal will be moral or constitutional. The developing countries are more vulnerable than others.

These statements were pointed out here because this issue is frequent in reliable international newspapers [31]. When rulers authorize laws that allow the excessive use of pesticides in the food of people who will die getting cancer that will be fine because it doesn't threaten the power or wealth of rulers or plutocrats. That will only represent fewer people competing for power or money. Therefore, most countries that prohibit (or deny) smart economy are the same ones that authorize the use of agrochemicals. Curious, isn't it? What about cancer? Why is there no cure?

In general, taxes must be fair to enable the state administration activity in a manner proportional to maintaining economic and educational growth. But taxation isn't allowed to being so high so as to make the human survival unfeasible. And when we have reports indicating that there are 275 million people living in extreme poverty in Central, Southern, and West Africa, West Asia, Latin America, and the Caribbean, it is possible stating that there is a lack of investments about generating income and education (source: United Nation's report: World Economic Situation and Prospects 2018 [32] from Department of Economic and Social Affairs - UN/ DESA; the United Nations Conference on Trade and Development—UNCTAD).

But the gap of investments isn't due to the low tax collection. In contrast, there are developing countries with profitable economic activity and with coffers full of tax money paid by citizens. But corruption and laundering money are responsible for poverty, hunger, and lack of professional opportunity. These added factors weaken democracy and create fertile ground for dictatorship.

For these reasons, the author argues that the growth of the economic flow isn't resolved by obligating citizens paying more and more taxes. There are already too many taxes in the world. The time now is to maximize these features.

This is the road of capital flow flight that most people don't see. But by far the biggest chunk of outflows has to do with unrecorded—and usually illicit—capital flight. US-based global financial integrity (GFI) calculates that developing countries have lost a total of $13.4tn through unrecorded capital flight since 1980 [33]. Most of these unrecorded outflows take place through the international trade system. Basically, corporations—foreign and domestic alike—report false prices on their trade invoices in order to spirit money out of developing countries directly into tax havens and secrecy jurisdictions, a practice known as "trade misinvoicing." Usually the goal is to evade taxes, but sometimes this practice is used to launder money or circumvent capital controls. In 2012, developing countries lost $700bn through trade misinvoicing, which outstripped aid receipts that year by a factor of five.

Alongside, taxes are captured in a larger proportion of the poorest people when expected otherwise. Mathematical calculations distort the principles of social equality which establishes the obligation of rulers to create fair and progressive taxes. As the rich become superrich because they pay lower taxes.

In the United States, the IRS found that as you go from being merely wealthy (the 1%) to super-duper wealthy (the 0.001%), your average federal income tax rate actually goes down. In other words, the progressivity of the federal income tax starts to fall apart at the upper reaches of the income distribution. Take a look [34]:

These issues aren't far from the technological field as smart economy uses blockchain technology and cryptocurrencies to free humanity from this cycle that violates human dignity and threatens human survival.

Paying taxes is necessary to generate benefits that will be shared between states and citizens, and paying just taxes is a real action to establish the principle of human dignity.

The Organisation for Economic Co-operation and Development (OECD) is worried about this scenario. Some countries are establishing the excessive taxation; double taxation principle is opposed by the OECD because it represents a practice potentially damaging the world economy and is unconstitutional. The method for elimination of double taxation is placed in Article 23 A and 23 B in OECD Model Tax Convention [35].

Double taxation is a principle about tax law referring to the income tax. It means the tax paid twice on the same source of income earned. It occurs when income is taxed at both the corporate and personal levels. Double taxation also occurs in international trade when the same income is taxed in two different countries.

This has been balanced by international agreements such as the Base Erosion and Profit Shifting (BEPS) Convention to Eliminate Double Taxation in Relation to Taxes on Income and Prevent Tax Evasion and Elision [36].

For the authors, income tax sounds like the tax from the feudal monarchy and double taxation will be set up when governments require tax revenue and, at the same time, are taxing products bought, supplied, and transported as well.

In most analyses of income distribution, we are asking for more reasonable sense to establishing taxes to the rich and poor people. The idea of greater income equality and a more even distribution of wealth reiterates the importance of a progressive income tax structure. It also highlights the economically sensible nature of targeted welfare assistance to those on lower incomes and a tightening of payments away from high-income earners.

This is where the current tax breaks to very wealthy people superannuants and the business sector needs to be radically overhauled. Not only will changing policies in these areas enhance economic growth and see a structural lowering in the unemployment rate, but they also have the other benefit of being fair, decent, and compassionate.

The authors are arguing that citizens should pay taxes only on consumer goods and services and not on income from the labor force: because it taxes the same source several times. Income from work has a noble social function because it is shared with society through taxes paid when the citizen buys products or uses services. Only income that is spared for short-term financial speculation should be directly taxed on income.

Throughout smart economy and the new business opportunities emerging in cloud environment allow us to think that it is possible to paying for proportional and fair taxes on income because there is no intermediaries and it eliminates the legal nexus to compose the tax liability directly on the income. And the public coffers will become full much more with the investments from stronger consumers through the consumption and investments that will become greater.

Smart economy is imperative for holding growing economic flow.

9 Sovereignty and Border Protection Requiring Inclusive Standards

From conferences and newspaper the authors brought us some interesting data. In information collected from 2017: 1.6 billion adults have a mobile phone and they can use this technology [37]. According to World Bank, 1.7 billion people are still unbanked [38] and countries with high mobile money account ownership have less gender inequality. In a world where the statistics is showing us that 1.6 billion adults have a cell phone but don't have access to a checking account this is able as an evidence about the statistical relationship which means that many people are excluded from the banking system because they create rules that exclude portions of the population assuming that this people are unprofitable for bank system. This behavior reflects on the world economy because excluding people from the payment system means restricting access to free trade.

The policy established by the banks is far from complying with the corporate principle of the company's social function and the humanitarian principle of the social function of property that requires every economic activity to generate jobs and to be nondiscriminatory. Thus, the classic banks have great influence and have caused damage to the world economy: it increases poverty and causes social inequalities [39].

Theory of disruptive refers to the socioeconomic phenomenon we are experiencing in this age. If people find themselves oppressed and unable to escape from hunger, they will manifest themselves anyway revolutionary. People in a natural and disorganized way will respond to the aggressive environment they inhabit as a way of survival. Blockchain technology emerged and smart economy immediately became the door to survival. This is just and legitimate. This is why the initial coin offering (ICOs) specializes through investments only from society, without participation of state investments. Besides this doesn't need to become a rule.

The legal concept created by the author, Romaoli, for disruptive theory is as follows:

Disruptive theory: It regards the process of social and business survival while introducing in the same time a foreign element to specific industry modifying the behavior of the whole phases of the sector: production processes, marketing, supply and demand policies, compliance, laws, and social and governmental behavior. It isn't a procedure driven only by competitive techniques to free commerce. Indeed, it means a procedure that reverses the order of power in that whole industry through technological practices with broader and effective communication in the intelligent environment. There is a balance between supply and demand. In fact it is a "leverage process" where social value is added.

Democratic power must recognize the legality and social function to smart economy and establish the necessary security to enable survival without changing the legal nature of the virtual economy with blockchain and cryptocurrencies.

Capital increases arising from the sale of stock certificate (when listed on the responsible bodies to control) and the capital gains on the disposal of cryptocurrencies acquired as a long-term investment and not intended to trade must be taxed under different rules.

The financial life of the ordinary citizen is a different right from the right that is available to companies when they are acquiring income from a trade or service such as securities and company shares. In this case, it must be reported in the tax return and taxed at the applicable rates because it impacts all members in trade society and doesn't just companies in cloud environment or virtual economy.

The ordinary citizen can't compete in the same level with companies. They don't enjoy legal instruments or tools comparable to those of corporation owners of legal personality to compete. This is just as states already guarantee that people are free to seek the cure of their illnesses by acquiring medicines and treatments that may not be available in the country. This is human dignity. No one can prevent the human being from doing what is necessary for the preservation of life. In the same reasoning and by equality of measure, money is equated with the remedy when in the proportion of guaranteeing the dignified life.

For this reason, the financial system addressed in smart economy enjoys humanitarian protection, has humanitarian value and social function, and must have different legal and tax treatment. This doesn't remove the monetary portion of governments to carry out the functions of public administration. In the other way, this increases the country's economy by generating more income to the states collected from taxes because a strong consumer buys more, circulates the currency, and heats up the economy.

Soon we will have questions about the smart economy to be resolved in the courts and therefore all the professionals that work in this industry need to speak the same legal language. In this way, they will be released themselves of crimes such as scams, frauds, laundering money, false currency, tax evasion, civil liabilities for material, and moral damages as well as violations of the consumer code.

This is such a crucial point that Malta, through Technology Arrangements Service Bill (TAS Bill), has set itself the obligation to draw up a law to establish the concept of smart contract before establishing the rules of responsibility.

The technical concept about smart contracts in 1994 is a computer protocol designed to facilitate, verify, or digitally enforce the negotiation or execution of a contract. Smart contracts enable the performance of trusted transactions without third parties [40]. But this concept isn't enough to be a legal technique. Romaoli, the author, has developed a broad legal concept that it is more appropriate to the current smart economy:

Smart contract: It means a decentralized management behaving like a mathematic ratio in self-executing contracts after being converted into codes with predefined business rules from computational code. In an effective sense, it is a obligation placed to machine to deliver rules from a hypothetic fact to a consequential fact under a "code interpreted in stricto sensu." In the end, it produces the individual and concrete legal norm valid between the contracting parties (NIC). Legal nature encompasses technological resources informed by AI and IoT establishing signs to be interpreted through the learning function of the machine to establish the applicable reality in prescriptive legal language. The intrinsic relation to the smart contract, blockchain, and AI tools represents the duty to be, that is, establishes the base proposition describing a hypothetical business event carrying the base proposition of the legal business relational to the consequent effect.

In addition, the legislative body from Malta Island was concerned about the validity and questioning about smart contract to the future. The reason is because smart contracts are differed from the classics as being self-executing and it works without additional third-party intervention. The system of positive laws was omissive about contracts in the intelligent format and currently we have led with IoT inserting elements to contract analyzing. Therefore, the laws that establish the concept and ways it will be applicable bring a greater degree of legal certainty to the intelligent contracts. Just illustrating the text, in Malta Island there was only Electronic Commerce Act (Chapter 426 of the Laws of Malta) stating that a contract would not be invalid simply because it is in the "electronic" form; it was proposed that a similar provision be introduced to ensure that a contract is not invalidated simply because of its smart format.

One of the most unique characteristics of blockchain is its decentralized quality of performance, which is shared among all parts of the network, thus eliminating the involvement of third party or intermediaries. There is a need to shift countries and cities from a government agency-centric model to citizen-centric model. This feature is particularly useful because it avoids the chances of any process conflict and saves time as well. While blockchains have their own set of issues that still need to be resolved, they offer faster, cheaper, and more efficient options compared to traditional systems. Because of this, even banks and government organizations are turning to blockchains these days.

Building the regulatory system is really important and should be in line with international law. Taxation has been a worldwide knot and in that respect Malta Island has dropped out of the way by building its regulatory framework.

Financial transactions and payments with blockchain are incorruptible and 100% auditable. The entries are intact and immutable and these are keeping laundering money away from legal business and do not feed corruption. Failures about enter data register are only possible to be manipulated by the human at first entry. However, the decentralized payment system is auditable and this would soon be discovered by the consensus. This feature discourages laundering money.

Smart contracts don't require intermediary as an outsourced facilitator. The consequence, essentially, is giving to user full control of the agreement. Blockchain acts as a mediator by loading the data/signs applicable to trading. It establishes a high trust between the parties and allows the reliable exchange of information. No one can steal or lose any of their documents because they are encrypted and stored securely in a shared and secure ledger. Also, you don't have to trust the people you are dealing with or expect them to trust you, because the impartial system of smart contracts replaces trust.

However, establishing inclusive access standards also involves thinking about the legal personality to the smart contracts. Many smart contracts will be hosted in the structure of a company already consolidated in a traditional way and the registration of this contract will follow this pattern. But this isn't always the rule. It happens that there are many smart contract manager businesses on cloud environment and therefore without a physical property structure. This could result in transactions occurring without an appropriate "legal entity" as the counterparty. Therefore, there is the need for a specific legal structure with requirements different from those in the classical register itself given the complexity of the technology involved.

For the authors, the emerging economy calls for a regulatory framework with inclusive standards in order to allow broad access in a nondiscriminatory and egalitarian way that a country cannot outgrow it. The compliance will focus on equalizing the individual benefits that are shared that develop the economic flow in a secure way.

Klaus Tilmes from the World Bank Group [41] suggests to realize this technological aspiration; we need to engage governments and people, coordinate development partners, and mobilize the private sector to:

1. Build: Develop the foundational building blocks for sustainable, technology-led economies
2. Boost: Expand the capacity of people and institutions to thrive in a resilient society in the face of disruption
3. Broker: Harness disruptive technology, data, and expertise to solve development challenges and manage risks

FATCA LAW and GDPR represent a legal control system for LAUNDERING MONEY. Of course it is positive. In the other hand, it has made it difficult for developing countries to access income because they are excluded from the system of bank payments for financial activities in case developing countries doesn't show up documents or reports in the same standard that was established by FATCA LAW or GDPR. In this case, automatically the source of income into transaction becomes subject to investigation about laundering money. Anyway, from a lawful criminal point of view, it isn't a right way to impose barriers taking as a premise the failure to fully comply with the requirements of these laws isn't the same as practicing crimes. This has represented trade barriers for some countries.

Going through this scenario, either the smart economy or just the financial system that communicates within the blockchain with the cryptocurrencies is the way for developing countries or small economies to access income through new professional positions, dignified and humanitarian taxation, and free financial autonomy from suspicious blockages or imposed by international banking regulations.

For the authors the typical man-machine interaction will make man better than he is today. AI will improve the way a man makes decisions in all industries because it will increase access to multiple information in a short time. This will require a great educational and technological development of the people.

For example, in medicine, AI will not provide cure for spinal cord injuries, but will provide patients with the same healing effect. That means a more dignified life.

Human reality will be interpreted by the machine that has been previously analyzed and fed by man with carefully studied information. The result will be smart contracts with errors reduced to zero or almost zero and no bureaucracy.

Note that this is a very high level of human development that poor countries don't have access to. Without proper education and policies, they will not have access to wealth and will be poorer. The result may be a multitude of hungry and desperate people. There may be motivation for wars or invasion of frontiers.

Thus the typical human-machine interaction will only be positive if richer countries develop policies to help poorer countries getting access to work and wealth. The border protection isn't through physical obstacle. The borders will be protected with virtual interaction and technological advance.

There is a common perception that the rulers agree with this dynamic relationship by verifying that there is no power without money. Even the hardest rulers and dictators have realized this and invest in diplomatic relations, and political, administrative, and social infrastructure. War generates weak economy and power only matters when correlated to economic power. There is a joint effort to invest in disruptive technologies to strengthen the world economy and this way holding the homeland security.

The shift in border security advancing from smart borders to virtual borders requires significant economic and physical infrastructure investments at or near the border in order to work as envisioned. International cooperation can also reduce the overall costs of necessary infrastructure. However, international cooperation in joint border infrastructure development and joint inspections may be too controversial politically in the immediate future when issues about the minimum existential to people are ignored. The upshot: significant economic and political barriers to implementing the smart border concept will bring worst consequences or rebound effect about sovereignty.

The stability of the citizen has layers that overlap: food, health, legal, social, and healthy environment. And economics is the armor that holds stability because it protects and provides a means of realizing the layers of stability. Politics matters only insofar as it serves to structure and maintain stability.

Physical borders will be protected only when human security is established. Globalization has broadened the border horizon by establishing virtual limits and requires governments to share responsibility for the security of mankind as technology promotes rapid industrial growth and education in developing countries cannot keep pace. Young people in emerging economies are forever playing catchup as a lack of foundational knowledge means advancing skill knowledge is moving further out of reach; already educated Western citizens are able to quickly reskill and work with machines. The partnerships between countries are required to promote equality.

Also Romaoli created the legal concept to worldwide responsible security:

The security of individuals maybe doesn't derive from that country safety. From my point of view, human security isn't placed just in military force that it means the late-stage and emergency action. Human security is placed in economy that can establish dignified life and it will drive reduction of civil and military conflicts because there is no motivation. Safety should be seen as emerging from the conditions of everyday life: food, education, housing, employment, health, and public safety. This is important because it means all conditions that generate stability and high level of freedom in maintaining individual safety. This framework along the social life will externalize the solidarity in a global chain and in a friendly environment able to establish the peace. My concept to human safety was extracted from article III of the Declaration of Human Rights, 1948, as a way to close framework to stablish peace from economy.

Blockchain technology combined with IoT, AI, and cryptocurrency divided society into new lines. However in the past, the separation was only by the wealth or elites from the academic industry that disputed the power by the technology. The shift was significant because it brought people closer to the sense of community. These elites are not just the rich but also groups of citizens with transnational interests and identities that often have common points of interest with their peers in other countries, whether industrialized or developing. Physical barriers no longer represent the borders of a sovereign country. Modern sovereignty is maintained by the inclusion of its citizens in the global world to access the wealth and assurance of promoting the same right for people residing outside their borders in a sense of global community.

Smart borders are not just a matter of deploying hardware, software, and soldiers. They require international cooperation. Existing smart border agreements are required to lay out an agenda for extensive and solidarity international cooperation, but even more cooperation will be necessary to collect the necessary data for the smart border concept to work in practice.

After all, countries need to agree about solidarity responsibility for human security. Then, in the content that the authors brought to the text, it is concluded that this agreement will go through economic efforts focused on partnerships to develop the technology as a human right because it will accelerate the economic flow and bring more satisfaction to humanity's life. If nothing is done in solidarity, the world will collapse long before 2030.

10   Conclusion and the One-Off Discussions

When we are talking about technology it means not only physical equipment—including infrastructure and installations (so-called artifacts), but also the knowledge, techniques, and skills that surround its deployment and use. These in turn form part of a broader technological "regime" or infrastructure that supports innovation and the ability for one technology to build on or link to another.

Generally, researches that address technology keep the focus on development and technical methods for creating, modifying, or improving sharing, use, and operativity.

The author innovates the research methods through a methodological cut under the legal, historical, technical, and socioeconomic aspect.

The research sources were useful for extracting data that were analyzed and together forming the structure that makes technological innovation with AI, big data, and the digital economy with blockchain and cryptocurrency survive the time to enforce the 2030 Agenda for Sustainable Development by United Nations to end hunger and poverty in the short time of 15 years.

The author concludes that the ANNUAL PARTICIPATIVE BUDGET PLAN strengthen instruments for popular participation in governmental decisions. The utility to establish the technologies as the Sixth Dimension of Human Rights is the treatment such as focal point to acquires investments to sustainable development. The results will drive the took-decision to maximize the tax application since fundamental rights is the base to allocate the portions of investments to public services.

So, after researches and observation, the author assures that the technology is raising the fundamental rights from fifth to sixth dimension. Thus, it points the technology specially artificial intelligence, big data, and digital economy with blockchain and cryptocurrency in a priority place to annual participative budget plan as a form to maximize the tax application thanks to transformative power.

The research is applicable as social, economic, and humanitarian policies and also as a doctrine to guide the regulatory world framework by governments, civil society, private companies, agencies, and nongovernmental organizations.

Since 2008 and to the present day there is still no inclusive standard of international norms to regulate the digital economy or use of technologies such as AI, IoT, and big data for social, humanitarian, environmental, or health application.

For this reason, fees, copyright, and royalties still require adjustment laws for the international community.

Have no doubts that interactions between human and machine means the future of work. Advances in artificial intelligence, robotics, and deep learning change how we work and jobs that were once widely viewed as safe are increasingly being automated, increasing migration and forced displacement. An augmented labor force emerges that marries the strength of human and machine. The localization is developing as additive manufacturing impacts supply chains, immigration, and urbanization.

Establishing that human civilization arrived to the 6th dimension of human rights goes further than building a humanitarian legal thesis. This means embodying the goals of the Declaration of Human Rights written in 1948 by the United Nations.

This adoption allows us to create a sustainable world along the lines of the 2030 Agenda for Sustainable Development by the United Nations.

The author is addressing the technology under the context of development, transfer, adoption, and dissemination and they are very thoughtful about the integration of scientific as an important element of policies and programs to manage natural resources in an environmentally and economically sustainable and culturally appropriate manner embodying governments, civil societies, private companies, and nongovernmental organizations to make possible that developing countries acquire the benefits from technological innovation. Thus, in a governmental responsibility for human security everybody will establish the sustainable world.

However, laundering money and frauds are pointed out as reasons for the emptying of public coffers and it makes governments to demand increasingly high rates from population to restore the amount that was gone away illegally. Therefore, these factors are jeopardizing human survival that it makes people working more and more to pay high taxes.

On the other hand, the population is growing fast every year without sufficient jobs to all. So, taxes is a real obligation that people cannot escape and they will pay even when they haven't got jobs. In the same way, the need for survival has brought conflict to the territorial borders.

It is a reason for governments to share the responsibility of human security. In addition to the gaps associated with the divide between developed and developing countries, equally important "divides" must be addressed to ensure that technology becomes an effective and equitable means to attain socially and ecologically sustainable development

So, increasing the economy is a urgent necessity of mankind and investments in technology can make it faster.

However, governments need to acquire taxes to manage its activities.

Watching the landscape that authors report, there arise three problems in humanity survival:

1. Increasing the economy and creating jobs is a urgent necessity and it must be done now and fast.
2. High tax is jeopardizing the human survival.
3. Governments must manage the state activities in the same time that they assume the responsibility for human security in a global field.

Which is the solution?

The technological innovation is the solution. However, governments need to accept this to apply investments in technology focused in improving the economy.

How the governments will make it?

Is required from countries to collectively discuss and to address the tax challenges arising from digitalisation, through mutual consensus as well any solutions should be long-term and have broad adoption by countries to allow for seamless application for business guaranteeing the Certainty and predictability.

The digitals companies have as attributes the ITINERANCY, MULTIPLICITY and VOLATILITY of the activities and for that reason to adopt a criterion for the unique tax nexus is not possible. To build this harmony in the Inclusive Regulatory will be possible if adding one more legal concept about Permanent Establishment: the "DIGITAL PERMANENT ESTABLISHMENT". Beside that, is mandatory think about to build the HYBRID THEORY TO THE TAX LEGAL NEXUS since it isn't possible to adapt the digital economy to international economic system in a harmoniously way without new premises to aims the legislative perfection and that complies with the principles: neutrality, non - double taxation, administrative's efficiency and enforceability, certainty and predictability, information security, flexibility, non-discrimination and dignity of the human being.

Creating a Social Tax in minimum limits will establish a balance between the country where the Income Tax will be paid and the country where there is the relevant economic presence. It is a compliance measure that allows countries with technological capacity to invest in developing countries generating income, jobs, education and economic improvement.

Developing countries need to adapt and to modernize their methods about tax collection. There is a need to discuss technology as FUNDAMENTAL RIGHT in the face of increasing technological innovations and impacts on the public finances.

So, after researches and observation, the author assures that the technology is raising the fundamental rights from fifth to sixth dimension. Thus, it points the technology specially artificial intelligence, big data, and digital economy with blockchain and cryptocurrency in a priority place to annual participative budget plan as a form to maximize the tax application thanks to transformative power.

Besides that, the annual budget plan and the participatory budget are part of the democratic system of governments where the governors and citizens define the destination of the money collected with taxes. It means investments in state administration and social programs.

Defining technology as a human rights dimension is to ensure that investments in technology will come up because it is the fundamental right required for survival. It is the reason why investments in technology may be demanded.

However, the research used a lot of statistical data to conclude that mankind pays too many taxes. That tax collection is diluted in corruption and money laundering.

Therefore, establishing the regulatory framework to smart economy is urgent because the economy with blockchain and cryptocurrency can release the mankind from corrupt cycle to reduce poverty and increase economic flow.

The government's formula for raising taxes to solve the problems about public administration is old, doesn't work, and jeopardizes the survival of humanity.

The problems that smart economy can contain were not ignored by the author who believes that it was possible to solve through investments in the regulation.

The time is to maximize tax money and don't create more taxes. It will accelerate the growth of humanity that is possible through technology.

But technology will be useless if the rulers don't accept it. Partnerships between civil society, the private sector, and governments are a faster way to allow access to education in developing countries.

All the measures that the author presented create a sustainable, harmonious, and safe world.

In parallel, new jobs requiring socio-emotional skills, curiosity, and entrepreneurship are emerging.

Thus, AI, cryptocurrency, and blockchain technology isn't a threat to humanity. This brings a better and more dignified life. However, the human being will never be able to compete with the machine. Then, we should change the way we educate children and adolescents to develop capacities like communication, partnerships, and skills to deal with conflicts and make solutions suitable.

We should prepare children to acquire principles and ethics in harmony with the sustainable world to establish the sustainable work satisfaction like one piece of dignity life.

References

1. Andrea Romaoli Garcia, AI and digital economy with Blockchain and Cryptocurrency like the 6th Dimension of Human Rights to border defense and a sustainable world. Panel: Blockchain for Social and Humanitarian Applications. IGF 2018: The Internet of Trust—Conference. Paris: UNESCO. November 13, 2018. Available http://www.intgovforum.org/multilingual/content/igf-2018-day-2-salle-vii-ws227-blockchain-for-social-and-humanitarian-applications. Accessed 10 Jan 2019
2. Andrea Romaoli Garcia, O desafio tecnológico para o pacto global em 2030: Blockchain, AI e defesa de fronteiras. São Paulo, Brazil: Webbr 2018 Conference, 4 October 2018. Available https://conferenciaweb.w3c.br/. Accessed 14 Oct 2018
3. J. Anderson, L. Rainie, A. Luchsinger, *Artificial Intelligence and the Future of Humans* (Pew Research Center: Elon University, Elon, NC). Dec 2018. Available http://www.elon.edu/e-web/imagining/surveys/2018_survey/AI_and_the_Future_of_Humans.xhtml. Accessed 14 Oct 2018
4. J.-J. Rousseau, *O Contrato Social. Tradução de Ciro Mioranza*, 2nd edn. (Escala, São Paulo, 2008)
5. F. Ulrich, *Bitcoin: a moeda na era digital*, 1st edn. rev. edn. (Instituto Ludwig Von Mises, São Paulo, 2014)
6. J.A. Tucker, F. Ulrich, *Bitcoin: a moeda na era digital*, 1st edn. rev. edn. (Instituto Ludwig Von Mises, São Paulo, 2014)
7. Jon Matonis, Bitcoin prevents monetary tyranny. Forbes, 2012. Available https://www.forbes.com/sites/jonmatonis/2012/10/04/bitcoin-prevents-monetarytyranny/#739963a32e8f. Accessed 10 Apr 2018.
8. Pinar Ardic, M. Heimann, N. Mylenko, Access to financial services and the financial inclusion—agenda around the world, in *Policy Research Working Paper*. World Bank Financial and Private Sector Development Consultative Group to Assist the Poor, 2011. Available https://openknowledge.worldbank.org/bitstream/handle/10986/3310/wps5537. Accessed 9 Apr 2018
9. E. Casalicchio, M. Caselli, A. Coletta, I.N. Fovino, DNS as critical infrastructure, the energy system case study. Int. J. Crit. Infrastruct. Indersci. Publ. **9**(1/2), 111–129 (2013)
10. Li Xia. Xinhuanet. China to invest multi-billion dollars to develop digital economy. Available http://www.xinhuanet.com/english/2018-09/19/c_137478730.htm. Accessed 21 Sep 2018
11. United Nations, Universal Declaration of Human Rights, 1948. Available at http://www.un.org/en/universal-declaration-human-rights/. Accessed 23 Nov 2018
12. OECD, 1948. Available http://www.oecd.org/. Accessed 23 Nov 2018
13. Esquerda Net, *EUA: Duas mil crianças migrantes enjauladas e separadas dos seus pais* (Carta Maior Newspaper, Brazil, 2018). Available https://www.cartamaior.com.br/?/Editoria/Antifascismo/EUA-Duas-mil-criancas-migrantes-enjauladas-e-separadas-dos-seus-pais/47/40655. Accessed 22 Nov 2018
14. O. Filtro, *Pelo menos 40 imigrantes morrem em barco no mar Mediterrâneo* (Epoca Newspaper, Brazil, 2015). Available https://epoca.globo.com/tempo/filtro/noticia/2015/08/pelo-menos-40-imigrantes-morrem-em-barco-no-mar-mediterraneo.html. Accessed 22 Nov 2018
15. P. de Barros Carvalho, *Direito Tributário Linguagem e Método*, 6th edn. (Noeses, São Paulo, Brazil, 2015)
16. T. Alex, *China to Enforce Regulation for Blockchain Companies in February* (CCN News, China, 2019). Available https://www.ccn.com/china-to-implement-regulation-for-blockchain-companies-in-february/. Accessed 11 Jan 2019
17. M. Doran, *A Forensic Look at Bitcoin Cryptocurrency* (Sans Institute, Estados Unidos, 2015). Available https://www.sans.org/reading-room/whitepapers/forensics/forensic-bitcoin-cryptocurrency-36437. Accessed 10 Apr 2018
18. N. Bobbio, *Liberalismo e Democracia* (São Paulo, Brasiliense, 2005)
19. P. Lenza, *Direito Constitucional esquematizado*, 13. ed. rev. edn. (Saraiva, São Paulo, 2009), pp. 669–675. atual. e ampl

20. Esteban Ortiz-Ospina, Max Roser, Taxation, Published online at Our WorldInData.org. Available https://ourworldindata.org/taxation. Accessed 18 Oct 2018
21. Andrea Romaoli Garcia. O desafio tecnológico para o pacto global em 2030: Blockchain, AI e defesa de fronteiras. São Paulo, Brazil: Webbr 2018 Conference; 4 Oct 2018. Available https://conferenciaweb.w3c.br/. Accessed 14 Oct 2018
22. K. Annan, *Technology Can Improve the State of the World* (MIT, Massachusetts, 2015). Available https://www.kofiannanfoundation.org/annan-work/how-technology-can-improve-the-state-of-the-world/. Accessed 10 Oct 2018
23. Sam Goundar. What is the Potential Impact of Using Mobile Devices in Education? ResearchGate. PDF available https://www.researchgate.net/publication/268337152_What_is_the_Potential_Impact_of_Using_Mobile_Devices_in_Education. Accessed 10 Oct 2018
24. Mauro Azeredo. World Bank. Digital Development: The World Bank provides knowledge and financing to help close the global digital divide, and make sure countries can take full advantage of the ongoing Digital Development revolution. 27 Sep 2017. Available http://www.worldbank.org/en/topic/digitaldevelopment/overview. Accessed 14 Oct 2018
25. United Nations Headquarters in New York, Transforming our world: the 2030 Agenda for Sustainable Development. 25–27 Sep 2015. Available https://sustainabledevelopment.un.org/post2015/transformingourworld. Accessed 14 Oct 2018
26. Jessica T. Mathew, The Power Shift: The Rise of Global Civil Society. Foreign Affairs Magazine, 1997. Available https://www.foreignaffairs.com/articles/1997-01-01/power-shift. Accessed 10 Oct 2018
27. United Nations, *World Population Projected to Reach 9.8 Billion in 2050, and 11.2 Billion in 2100* (Department of Economic and Social Affairs, New York, 2017). Available https://www.un.org/development/desa/en/news/population/world-population-prospects-2017.html. Accessed 18 Oct 2018
28. M.A. Boden, Creativity and artificial intelligence. Artif. Intell. J. Elsevier: England **103**, 347–356 (1998)
29. Terry Nguyen, West Virginia to offer mobile blockchain voting app for overseas voters in November election, The Washington Post Journal.10 Aug 2018. Available https://www.washingtonpost.com/technology/2018/08/10/west-virginia-pilots-mobile-blockchain-voting-app-overseas-voters-november-election/?utm_term=.e0333d21a489. Accessed 10 Oct 2018
30. Explaining Greece's Debt Crisis, New York Times Journal. 17 June 2016. Available https://www.nytimes.com/interactive/2016/business/international/greece-debt-crisis-euro.html. Accessed 14 Oct 2018
31. Paulo Prada. Why Brazil has a big appetite for risky pesticides. Reuters Investigates. 2 Apr 2015. Available https://www.reuters.com/investigates/special-report/brazil-pesticides/. Accessed 14 Oct 2018
32. United Nations, World Economic Situation and Prospects 2018, United Nations Publications. ISBN: 978-92-1-109177-9 e ISBN: 978-92-1-362882-9. PDF Available https://www.un.org/development/desa/dpad/wp-content/uploads/sites/45/publication/WESP2018_Full_Web-1.pdf. Accessed 18 Oct 2018
33. Jason Hickel, Aid in reverse: how poor countries develop rich countries, The Guardian Journal. 14 Jan 2017. Available https://www.theguardian.com/global-development-professionals-network/2017/jan/14/aid-in-reverse-how-poor-countries-develop-rich-countries. Accessed 14 Oct 2018
34. Christopher Ingraham, As the rich become super-rich, they pay lower taxes, For real, The Washington Post Journal, 4 June 2015. Available https://www.washingtonpost.com/news/wonk/wp/2015/06/04/as-the-rich-become-super-rich-they-pay-lower-taxes-for-real/?noredirect=on&utm_term=.a1b0e1e1aaa2. Accessed 14 Oct 2018
35. Khristine Kaddous. The European Union in International Organisations and Global Governance. Oxford and Portland: Oregon. 2015. pag. 261.
36. OECD, Multilateral Convention to Implement Tax Treaty Related Measures to Prevent BEPS, Available http://www.oecd.org/tax/treaties/multilateral-convention-to-implement-tax-treaty-related-measures-to-prevent-beps.htm. Accessed 19 Oct 2018

37. ITU-UN, ITU TELECOM WORLD 2018 conference: Durban, setembro 2018. Available https://telecomworld.itu.int/2018-event/smart-abc/. Accessed 18 Oct 2018
38. The World Bank, Financial Inclusion: Financial inclusion is a key enabler to reducing poverty and boosting prosperity, 2 Oct 2018. Available https://www.worldbank.org/en/topic/financialinclusion/overview. Accessed 18 Oct 2018
39. F.U. Coelho, *Manual de Direito Comercial: Direito de Empresa*, 18th edn. (Saraiva, São Paulo, 2007), p. 497
40. Nicholas J. Szabo, The George Washington University Law School, 1994
41. Klaus Tilmes, Naoto Kanehira, Perspectives on Disruptive Technologies and Forces, World Bank Group. may 10, 2018. Available http://pubdocs.worldbank.org/en/10808152600375 2002/051118-disruptive-technology-seminar-Tilmes-Klaus.pdf. Accessed 10 Oct 2018

Part II
Experiences

Chapter 7
Modelling of Traffic Load by the DataFromSky System in the Smart City Concept

V. Adamec, D. Herman, B. Schullerova, and M. Urbanek

1 Introduction

Transport has always been an integral part of society. Modern society couldn't exist today without constant transport of goods, products and information. However, it also has a negative impact such as air pollution and environmental pollution. This may cause damage to human health from serious illnesses to untimely death. The pollutants also affect vegetation and may cause reduction in agricultural production. They even cause damage to materials and buildings of historical significance. In recent years, the share of automobile transport in the air pollution has significantly grown, particularly in high-traffic urban areas. In 2015, there were more than 200,000 new passenger cars in the Czech Republic and for example in Germany more than 3 million [30].

The Sperling and Gordon study [37] assumes that by the year 2030, there will be about 2 trillion cars driven in the world. Transport brings risks to human health and the environment in the form of air pollution from exhaust fumes in the form of CO, NOx and volatile organic substances. What also increases the emission load in cities significantly is congestion. 70–80% of the European population currently lives in cities, where they also meet and move around. [1]. Air pollution, temperature increase, noise, accidents and immobility of the inhabitants are connected with increased untimely death rate and serious illnesses, especially in cities ([2, 3] [32]). According to Zhao and Zhu [4], about 30% of traffic congestion is caused when drivers are trying to find a free parking space. For example, the results of a study in Barcelona show that every day around a million vehicles spend 20 min searching for

V. Adamec (✉) · B. Schullerova · M. Urbanek
Brno University of Technology, Institute of Forensic Engineering, Brno, Czech Republic
e-mail: vladimir.adamec@usi.vutbr.cz

D. Herman
RCE Systems, Ltd., Brno, Czech Republic

© Springer Nature Switzerland AG 2020   135
N. V. M. Lopes (ed.), *Smart Governance for Cities: Perspectives and Experiences*,
EAI/Springer Innovations in Communication and Computing,
https://doi.org/10.1007/978-3-030-22070-9_7

a parking space, which produces 2400 tons of CO_2. The INFRAS study (2000) says that traffic congestion, pollution and road accidents in EU countries cost 502 trillion euro a year. Especially the economically quickly developing countries are trying to develop strategies for reducing air pollution not only in cities. Nieuwenhuijsen and Khreis [1] give an example, which monitors current strategies for introducing cities without cars and promotes more pedestrian and bicycle zones for a healthy city life. It compares these strategies with, for example, the cities of Hamburg, Madrid or Oslo, which have been implementing their plans in the public transport and personal transport is already partly limited here. It also mentions cities like Brussels, Copenhagen, Dublin, Paris or Bogota, where various measures have been taken to limit vehicular traffic by, for example, supporting cycling, offering benefits when using public transport or reducing the number of parking spaces in the city centres [38]. Other strategic measures include Carsharing, which, thanks to connecting with applications in smart mobile devices, allows an increase in efficiency and use of this service [5]. The same applies to using the so-called Mobypark applications, which allow sharing information about free parking spaces in cities, near public buildings, in hospitals, in public parking houses or in hotels [6].

The EU Commission is preparing a strategy for clean transport, which should become effective after 2020, and it is in compliance with the strategy proposal for low-emission transport. One of the significant impulses is the ever-increasing air pollution in cities and a high share (up to one-third) of road transport in the creation of greenhouse gases [39]. The total share of transport in the creation of greenhouse gases is around 23%, as shown by the EUROSTAT data [40].

Decreasing the emission load in cities is one of the main goals of the smart city concept [7, 8], which looks for tools which would not only help decrease the emissions and prevent them but also allow precise monitoring with fast and precise evaluation for specific sections and areas. Transport and infrastructure is, therefore, a promising field, in which the latest technologies are used aiming to ensure traffic flow, its reliability and the decrease in the emission load already mentioned.

Monitoring emissions is one of the necessary parts of the measures whose goal is to increase the quality of the lives of inhabitants not only in cities. In the Czech Republic, transport emissions are monitored based on either real measurements, calculation or modelling. This text briefly introduces these approaches.

2   Current Approaches to Monitoring the Emission Load from Traffic in Cities

To determine the share of mobile sources in air pollution, the MEFA 13 emission model is used in the Czech Republic. It allows the user to:

- Obtain fully automatic calculations of emissions for any number of line sources (road sections)
- Define their own vehicle fleet composition

- Obtain a calculation distinguishing personal, lightweight/heavyweight trucks and buses
- Obtain a calculation of emissions from intersection crossings
- Obtain a calculation of emissions for single vehicles

The calculations made by the model include emissions from cold starts, non-standard riding modes when crossing an intersection (the stop-and-go mode), load distribution among trucks and consideration of brake and tyre wear. The programme output is emission values for each pollutant specified by legislation [9]. Other examples include the Stationary Sources Modelling System Methodology (SYMOS'97), which is used for conducting dispersion studies for air quality assessment and which allows the following [36]:

- Calculation of air pollution caused by gaseous and dusty substances from point, line and areal sources
- Calculation of pollution from a large number of sources
- Determination of the pollution characteristics in a dense net of reference points and thus preparing the materials necessary for clear cartographic processing of the calculation results
- Accepting the statistical distribution of the direction and speed of wind related to the stability classes of the boundary layer of the atmosphere based on the classification by Bubník and Koldovský [36]

To model motor vehicle emissions, the EPA (Environmental Protection Agency) in the United States uses the MOVES system (Motor Vehicle Emission Simulator). It is a system which models emissions from mobile sources (especially motor vehicles on motorways) at the national and regional levels. The MOVES is able to simulate emission rating from the year 1999 to 2050. The agency provides the model completely free of charge under the GNU licence and that is why it is used in a lot of research studies [10].

The German company PTV Group is developing a set of programmes called the Vision Traffic Suite. It enables a computer simulation of traffic and transport planning including vehicle and pedestrian simulations. The set consists of the following programmes: PTV Visum—focused on macroscopic traffic modelling, PTV Vistro—focused on the analysis and simulation of the impact of traffic reorganisation, PTV Vissim—focused on microscopic traffic modelling and PTV Viswalk—focused on pedestrian simulation. Besides the traffic analysis, the programmes allow emission simulation which considers the driver's behaviour. The PTV Vissim can simulate not only motor vehicles but also public transport railway vehicles, pedestrians and cyclists [11] [28] (Fig. 7.1).

Another company dealing with traffic emission modelling is Emisia, which is, besides other things, developing two programmes: Sibyl and Copert. The simulation programme Sibyl is designed for assessing the impact of strategic plans. It allows prediction of the number of vehicles and emission values in a country until the year 2050. The Copert system uses data on the number of vehicles, their speed and distance travelled and on the outdoor environment parameters, using which it calculates the

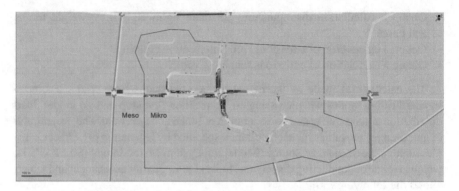

Fig. 7.1 Example of hybrid (meso + micro) simulation of traffic in PTV Vissim (source: [11])

Fig. 7.2 Example of street-level emission map with user interface (source: [12])

emissions and energy consumption of the region. The Copert programme development is administered by EEA (European Environment Agency) and it is developed for the official inventory of road transport emissions in countries falling under EEA. Besides this, the programme can also be used in research, scientific and academic applications. Copert models emissions from three basic operation modes: the thermally stabilised engine mode, the warm-up-phase mode ("cold-start" emissions) and the non-exhaust emissions (fuel vapours, brake and tyre wear). Like the American MOVES model, Copert is also freely available [12] (Fig. 7.2).

To determine the traffic emission load, the real driving emission (RDE) measuring may be used. It measures the pollutants produced by vehicles in real traffic.

To carry out the measurement, a portable emission measurement system (PEMS) is attached to a vehicle or placed inside it and a PEMS test is done in traffic on a random route in several modes: riding at a higher speed on a motorway, riding at a medium speed on out-of-town roads, riding at a low speed in a town and riding in hilly terrain. The test takes into consideration the influence of the surroundings (temperature, pressure), the load added to the vehicle and the effect of riding up or down a hill. The real route is monitored by a GPS device during the test. The result of this test is the determination of the conformity factor, which is the maximum emission limit set by the legislation, which also considers a measurement error. The effect of an error is more significant in PEMS than in laboratory tests [13, 14].

Transport Research Centre (CDV) and SEKO Brno deal with the measurement of vehicle emissions in real traffic using so-called e-trailer, which is attached to the vehicle with towing equipment. The measurement includes continuous monitoring of gaseous emission concentrations of oxygen, carbon monoxide, carbon dioxide, hydrocarbons and nitrogen oxides and also isokinetic sampling of particulate matter and gases into bags to be analysed consequently in a laboratory. The measurement is carried out in more riding cycles: predefined (standardised, custom), in traffic on a standard route and on a random route according to the PEMS procedure [13, 15] (Fig. 7.3).

Other possible ways of monitoring traffic emission load were researched at Konkuk University. They used cameras which were able to identify the number and type of vehicles passing through. Based on this data and information on the emission values of different vehicle types, they modelled the total amount of emissions in real time. The data is stored in a server database and it can be viewed using an Internet browser.

Thanks to the identification of the vehicle types, the amount of emissions produced by each type (cars, trucks and others) can be determined. In the research, four pollutants were measured: carbon monoxide (CO), nitrogen oxides (NO_x), particulate matter (PM_{10}) and volatile organic compounds (VOC).

To determine the traffic density, the type of vehicles passing through and their speed, traffic parameter estimate based on camera recordings was used. To obtain the estimate, traffic images were compared with a reference image without traffic and the traffic density and speed of vehicles were determined based on colour differences. Based on these, the number of vehicles was calculated. The total amount of emissions from one type of vehicle per hour is calculated using the following formula:

$$PE = N \cdot D \cdot C$$

where PE—amount of emissions, N—number of vehicles, D—distance travelled per hour and C—emission coefficient for the vehicle type.

The cameras used in this research use the wireless sensor network (WSN). They are linked to a storage unit using wireless technology. The camera recordings are saved in two forms: as JPG images (30 images taken per second) and as ASF video

Fig. 7.3 E-trailer (source: [13])

recordings. Besides the video recordings, traffic parameters in XML format are also recorded on the servers.

For comparison, another technology is introduced. It is based on monitoring emissions in areas with high traffic load or in so-called low-emission zones. An example is using a camera system enabling vehicle identification with an emission detector in real time (NOx, PM, CH, CO, etc.). The system identifies the vehicle based on their type (trucks, buses, vans, cars, etc.) [16]. However, the system has, like most other systems, its weaknesses regarding the vehicle identification, when the camera system is not able to distinguish whether it is diesel, petrol or another type of fuel.

3   Smart City Approaches to Monitoring and Assessment of Air Quality in Cities

In the sphere of air pollution, the smart city concept methodology primarily deals with preventing the production of emissions in various ways. In transport, they are especially the following:

- Restriction of entry and traffic jams in city centres (both for individual and freight transport)
- Increasing the number of trees along roads in cities
- Support for green corridors for cyclists and pedestrians
- Support for clean traffic especially in city centres
- Low-emission or no-emission zones
- Dynamically controlled intersections and navigation

In the sphere of energetics, they are the following:

- Monitoring the development of CO_2 emissions
- Restriction of local solid fuel furnaces in city centres
- Increasing the proportion of energy production from renewable sources
- Support for local production and local energy consumption
- Production of heat through municipal waste incineration

Smart technologies can be used not only to prevent but also to monitor pollution and subsequently manage the emission sources to improve the situation [17, 18].

3.1   Climo

In cooperation with Intel Corporation, the Robert Bosch GmbH company has developed a smart monitoring system called Climo, which enables fast and accurate measurement of air pollution. This system can simultaneously measure and save/share data on the main pollutants, which are solid particles, carbon monoxide, nitrogen oxides, sulphur dioxide and ozone. Besides that, the system also records the environment parameters such as temperature, relative humidity, pressure, pollen concentration, light and sound. Compared to other standard monitoring systems, Climo is smaller and easier to maintain but it monitors a much smaller area. Data collection and subsequent sharing are possible using wireless (Wi-Fi and 3G) or wired connection. Climo offers some advantages of Intel technology such as Cloud analysis, data management and visualisation software. The data measured may be used for better traffic flow management or to inform inhabitants with respiratory problems or allergies about unfavourable conditions in certain parts of the city. In 2018, the Climo monitoring system received the CES 2018 Innovation Award in the Smart Cities category [19].

3.2 Monitoring the Environment Pollution Using WSN

WSN, wireless sensor network, is a group of small autonomous independent sensors intended for monitoring the environment conditions. The information is passed using wireless interconnections and stored in the main collecting locations (servers). A newer WSN enables bidirectional connection and therefore the sensors may be managed individually. The sensors consist of several independent components: a communication element, memory, a sensor a controller and a power source. The communication element may be Bluetooth, ZigBee, LTE or Wi-Fi [20].

In most wireless sensor networks there must be at least one node which is only used for collecting and transferring data to be processed. When there are a large number of sensors, there may be a bad connection between the sensors and the node collecting the data, which could receive wrong or no information. To eliminate these problems, the Cluster method may be used, which means that a cluster of sensors pass information to the connecting nodes which share it with one common server [21].

WSN can be placed in public transport vehicles which move through a city and thus cover a large area designated for air pollution measurement. In addition to pollution measurement, the sensors can also monitor, for example, information about passengers or other smart transport parameters. To carry out the measurement in areas where wireless transfer is not possible, the sensors may be placed at bus stops, which collect the data into their own memory. Buses thus create intermediate nodes enabling data collection from the sensors when standing at the stop and subsequent sharing with the server using an LTE network [20].

3.3 The Conception of a Support System for Decision-Making Using IoT

IoT—Internet of Things—is a network of physical devices, vehicles, home appliances and other electronic equipment which allows these devices to connect, collect and exchange data. In the smart city projects, the Internet of Things is one of the essential parts. It constitutes the interconnection between the systems and subsystems of a smart city. Thanks to this it is possible to monitor the current situation in the city and respond dynamically. This mainly concerns monitoring the traffic and air pollution related to it.

The conception of a support system for decision-making in smart cities uses the Internet of Things to collect data from stationary and mobile sensors, information from traffic or individual cars and information on the weather. This data is modelled and then used in strategies for the reduction of pollution (detours, optimisation of signalling at intersections, parking regulation and prohibition of truck entry). The strategy may be compared with the previous situation and its efficiency may be established. To make predictions using modelling, it is important to determine the drivers' route selection logic so that it is in maximum accordance with reality when a new detour or closure is made. It is also necessary to specify the emission models of each vehicle and to estimate the atmospheric dispersion correctly [22].

3.4   BreezoMeter

The BreezoMeter Ltd. Company developed an application which displays a global map of air pollution. Besides displaying the air quality index, the application also contains the values of each pollutant, a forecast and history of the air quality index, allergen values and recommendations for persons with respiratory problems, children and sport activities. The application can be installed in a smartphone, where it monitors the user's movement through GPS and it warns them if they are in a zone with low air quality.

The main difference between the current approach to measurement and the BreezoMeter conception is that the current approach uses the national pollution monitoring network, which only displays data from the surroundings of the stationary measuring stations (AIM). BreezoMeter collects global information from several sources such as information on the weather, traffic flow, GPS and data collection from measuring stations. These so-called Big Data are gathered and verified on servers and then used for modelling a complete air quality index. The dispersive interpolation algorithms used for modelling are based on machine learning techniques. The BreezoMeter company offers interconnection of the application with the automotive sphere, smart cities, smart homes and others [23] (Fig. 7.4).

In accordance with the smart city concept, strategies which require demanding monitoring of the air quality are introduced in cities. The Smart Air Quality System is one example which is used in India. It evaluates data from the closed-circuit

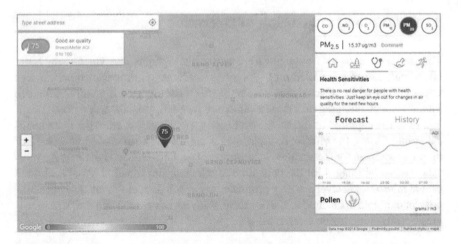

Fig. 7.4 BreezoMeter—the application interface (source: [23])

television (CCTV) such as the total number of vehicles and the number based on the type of vehicle occurring in a specific area and considering the temperature, speed or density is very important [24]. Short-term monitoring (in the range of hours) is possible thanks to software tools available in the form of a Web application which shows the development of the emission situation graphically and numerically [24, 25].

It is the possibility of short-term or continuous monitoring of the emission load from traffic in cities which allows more efficient traffic planning and control and thus dealing with undesirable situations in which traffic congestion or other collisions occur and the concentration of vehicles in the particular section increases. That is why it is beneficial to apply this measure already in the traffic planning stage, if possible before the renovation or construction of new roads, intersections, buildings, etc. A complex problem to deal with is the historic city centres and old built-up areas which were originally designed for completely different traffic capacities which are no longer sufficient. That is why measures such as diversion routes, entry of vehicles only with permission and reinforcement of the public transport system are taken. However, these measures are not sufficient in all cities. Therefore, there is also a possibility of monitoring the traffic online with immediate evaluation and if there is a connection with the traffic centre, traffic control is also possible, for example by changing the time intervals at traffic light intersections, traffic control by the police, changing the driving direction in some streets in critical daytime periods or banning the entry of vehicles in these time periods.

One possible way of implementing a software tool in the smart city concept is using online monitoring of the traffic and its analysis in selected sections which are, for example, often affected by high levels of traffic and traffic congestion. That is why the following text introduces the DataFromSky software tool [41], which was developed in the Czech Republic.

4   DataFromSky: Principle of Operation

This software uses data obtained from aerial photographs and videos which are taken with cameras installed on high-storey buildings, on drones or other devices with the view of the selected area. Extracting the trajectory data from an aerial video footage is a challenging task due to camera movement, lens distortion, visual variability of the captured traffic scene, occlusions, etc. However, from the mathematical point of view, the bird's-eye-view images are the most suitable input in case of monocular vision for the accurate localisation of ground targets. Therefore, the utilisation of aerial images for the collection of telemetry information about each road user at microscopic level is a very promising approach. It opens many new possibilities, for example in the field of traffic safety, studying of complicated traffic manoeuvres or estimating traffic emissions.

The system analyses and processes the data in two stages:

1. Georegistration: Establishing correspondence and mapping between video sequence frames and real-world coordinate system
2. Detection, localisation and tracking of objects of interest (vehicles) in the geo-registered video sequence

For the simplicity, the georegistration process assumes that the road surface is planar and consists of two steps: camera calibration procedure and selection of at least four points in the reference image of which the exact positions are known. These points are used for estimation of transformations that allows mapping a pixel from image space into coordinate system of an intersection. The precision of mapping is influenced by the following factors:

- Quality of camera calibration: undistortion coefficients/model
- Quality of image stabilisation: estimation of UAV movement between consecutive images
- The precision of geo-referenced points and their localisation in the image space
- Planarity of the road surface

Automatic (re-)identification of objects of interest and their tracking during their passage through an intersection is a very difficult task because of complex background, dramatic appearance variations, different light conditions, camera movements, etc. Recent very promising state-of-the-art methods are based on deep neural networks for both detection and tracking [31, 33, 35].

Object detection is a task of generating class labels with bounding boxes for one or more objects in an image. Convolution neural network for object detectors like Fast R-CNN and Faster R-CNN exploits the feature extractor output to propose and classify regions of interests. To improve the robustness and performance, the detection candidates are pre-filtered by expected area of road surface and results of moving object detection. The detections are then matched and tracked through the video sequence [29, 34].

Trackers for object tracking are typically trained entirely online. A standard approach samples patches near to the expected location of the tracked object. These patches are then used to adapt internal classifier which is used to evaluate patches from the consecutive frame to estimate a new location of the object. But this type of tracking is very computationally demanding and, therefore, too slow for multi-target scenes, i.e. with more than five objects. To overcome this issue, the new trackers utilise only the detections without any visual information, so-called tracking-by-detection approaches [27]. In other words, it is expected that the quality of the detector is adequate to mitigate the need for a strong tracking solution. In the first phase, a basic tracker connects very close detections into tracklets. In the next step, the tracklets are iteratively connected into the trajectories and/or with the existing trajectories based on the evaluation of each possible candidate solution using a score function. The score function considers distance in the time and space of the endpoints of the candidates, correspondence in the types of objects and prediction of their movements.

To eliminate localisation noise, the trajectories are filtered to form smoothed data based on vehicle kinematic models. The filtration process suppresses the noise of the detector in the localisation of an object in the image and, therefore, helps to dramatically increase the precision of speed and acceleration measurements.

5   Areas of Use: DataFromSky

The output of the analysis is a detailed telemetric information about every single participant of the given traffic flow, measured with high accuracy and frequency (in regard to the expanse of the monitored area), which is up to 25 Hz (depending on the frame rate of an input video). Another advantage of this approach is the fact that the observed participant of the traffic flow is not aware of being observed. Therefore, the distortion in data, which is common to other methods of gathering data from traffic, is not present there (as the UAV is practically not visible for drivers from a certain altitude). Traffic data with these characteristics allow for new possible areas of use in the field of traffic analysis.

These possible areas of use include:

- Motion pattern classification, abnormal motion pattern detection
- Traffic mining, prediction of non-recurrent short-term traffic patterns
- Developing next-generation driving behaviour models, trajectory clustering
- New perspective on merging behaviour—gap seeking instead of acceptance behaviour
- Evaluation of traffic simulation methods/models on real data, capacity estimations
- Traffic congestion studies, travel time, vehicle counting, trajectory-based analysis
- Road safety, pedestrian and cyclist safety, road–user interactions, smart analysis
- Driver decision-making and complex multi-actor interactions, traffic instabilities
- Empirical evidence of multi-anticipation (drivers have multiple leaders)
- Study of behaviour adaptation effect

The examples mentioned above regard only the reverse analysis of the acquired data sets, while the conclusiveness of the results derived from these datasets is, besides other factors, influenced by the size of the analysed sample and the measuring accuracy. As there are high technical demands of acquiring aerial data, alongside with the legislative limitations of using UAV, it is not simple to gather representative and sufficient-sized samples to make relevant conclusions in some research applications. In these cases, there is an interesting possibility of connection with advanced (traffic) simulators, while the valuable and information-rich data, acquired from the DataFromSky system, is used to calibrate simulated models of behaviour at the given analysed task to improve the operation of the simulator. The possibility of simulator calibration on the basis of smaller sample of information-rich data is a way to accelerate studies of traffic, while the information value of the consequent conclusions is not negatively influenced.

A brand new area of use of an aerial optical localisation system is the processing of traffic data in real time. As this functionality is now available, it makes new traffic applications, which require extreme mobility and a system with rapid application for acquiring traffic data, possible. It can be for example an adaption of signal plans of traffic lights according to the current traffic situation in case of an accident or other unexpected situations on the road. In the future, this solution could also be applied in the field of traffic control, including speed controls, control of a safe distance between vehicles, violations of rules of reserved traffic lanes and automatic detection of road rage (Fig. 7.5).

Apart from its use in traffic monitoring, the DataFromSky system is a tool sought after also for monitoring pedestrians and cyclists. This segment includes research of pedestrian transport, calibration of behaviour models of individuals in the crowd and even detection of dangerous and unusual or so-called anomalous behaviour of individuals. This approach to gathering data is, thanks to the rapid development of autonomous means of traffic, gradually being implemented also in research regarding examination of interactions between autonomous vehicles and other vehicles and/or pedestrians and cyclists. The following chapter includes numerous practical utilisations of the DataFromSky system.

Fig. 7.5 Illustration of a police drone—a system for traffic monitoring in real time (source: [41])

6 Examples and Outputs: The Application of the DataFromSky Software

As a standard use of analysis of aerial video data for traffic research, DataFromSky calculates the O/D matrix. For generating an O/D matrix from the extracted trajectories, it is necessary to define so-called virtual counting gate. This gate is able to reproduce even geometrically complicated crossroad traffic flows, as it is defined using polygonal curve. To gather even highly detailed information on the usage of the individual segments, gates can be differentiated for individual lanes on the road. When exporting the fundamental parameters of the traffic flow, it is possible to set up time granularity. With the first derivative of the current position in time, the velocity of the vehicle can be acquired. With the second derivative, acceleration of vehicles is then indirectly measured from the data. This tool, therefore, allows the user to easily acquire speed profiles and passage times in the measured traffic flow. It also offers a wide range of filters for selection of object of a certain type, selection of passages in the chosen direction, a certain time interval, etc. In the following figure, a lot of these functions and their outputs are illustrated (Fig. 7.6).

Traffic flow analysis from the viewpoint of security is yet another interesting application of this system. Telemetric data is used for identification of "conflict situations", which are detected on the basis of several indicators, such as time to collision (TTC) (using the model of vehicle motion—Ackerman steering), post-encroachment time (PET), time exposed time to collision (TET) and maximal deceleration. According to the number of these conflict situations and their distribution in individual sectors of the crossroad traffic flow, potentially dangerous spots can be identified and further analysed by a traffic expert. This approach, based on using telemetric data, is continuously developed (Fig. 7.7).

Fig. 7.6 Example of the output from DataFromSky—O/D matrix, speed measuring and object classification (source: [41] and COWI A/S)

Fig. 7.7 Example of detection of near accidents in the traffic flow based on TTC indicator (source: [41])

In a wide range of situations, DataFromSky can also be applied for calibration of models describing the behaviour of participants in the traffic flow. One example of such situation is research of decision processes of drivers when entering the highway from the merge lane, while several parameters are evaluated (time gap between two vehicles on the main road, speed difference, traffic intensity, etc.). The parameters which can be measured by the system also include GAP TIME and TIME TO FOLLOW. These parameters are used for calculation of the crossroad's capacity or calibration of models of driver's behaviour on roundabouts. DataFromSky is a useful tool also in projects concerning safety in the proximity of schools or nursery schools, where the effect of safety measures on the traffic (such as sleeping on the road) is compared before and after its implementation. This approach is, at the microscopic level, among the most promising possibilities of using real measurements from actual traffic flows for analysis (Fig. 7.8).

Another advantage of the software is the possibility of using the real data obtained for modelling and monitoring the emission load in specific areas. They are not only intersections but also other sections loaded with traffic which are often problematic in terms of the emergence of traffic congestion, etc. The selection of the way of monitoring using the DFS software is also important with regard to the possibilities of the area monitored. Specifically, in some countries and at some places, the selection of drone technologies is currently limited by legislation for safety reasons. That is why it is possible to select other monitoring alternatives such as mobile forklift devices and high-rise buildings. Another advantage is the fact that the system enables the monitoring of traffic even under impaired dispersion conditions or at night-time. It involves using infrared cameras, which are able to detect the individual objects of interest. The current effort is to model traffic situations and assess the emission load using real data by way of connection with software such as TIMIS or VISSIM [28]. One of the disadvantages of the DFS software is that, like the other systems mentioned in the previous chapters, it is not able to distinguish the vehicles based on the type of fuel they use (diesel, petrol, LPG, CNG, etc.).

Fig. 7.8 Example of utilisation of DataFromSky for a calibration simulation model for more precise capacity estimation (source: [41])

7   Conclusion

The software currently allows an analysis of real data which can be used for traffic control and planning not only in cities but also for any road section, intersection, parking place, etc. Its application may be useful in monitoring, analysing and evaluating the traffic, when information on vehicle counts, flow gap time, follow time and origin destination matrix is offered. It is also able to measure a macroscopic traffic flow characteristic at any point or region by analysing all vehicle trajectories in the place, speed and acceleration records for all vehicles tracked in a video. It is able to classify every tracked vehicle according to its visual appearance. Using the data mentioned above, application of the DataFromSky software to measuring the emission load in cities is currently being dealt with. The model area is the city of Brno. The research focuses on a possible connection of the software tool with the TIMIS software which is currently used [7]. Another possibility is extending the DataFromSky SW functions by modelling the emission load during the analysis of the real data obtained. Ensuring the compatibility with another software tool or carrying out direct calculation modelling increases the efficiency and the possibility of evaluating the immediate situation in real time in the section monitored [26]. The aim is to create a supporting tool for traffic control which is in accordance with the smart city concept and leads to reducing the emission load from traffic.

References

1. M.J. Nieuwenhuijsen, H. Khreis, Car free cities: pathway to healthy urban living. *Environ. Int.* **94**, 251–262 (2016). ISSN: 1550-1477
2. K. Bhalla, M. Shotten, et al., *Transport for Health: The Global Burden of Disease From Motorized Road Transport [online]* (World Bank Group, Washington, DC, 2014). [cit: 2018-09-10]. Available on the Internet: http://documents.worldbank.org/curated/en/2014/01/19308007/transport-health-global-burden-disease-motorized-road-transport

3. WHO, *Global Health Observatory* [online], 2015 [2018-09-02]. Available on the Internet: http://www.who.int/gho/road_safety/mortality/en/

4. H. Zhao, J. Zhu, Efficient data dissemination in urban VANETs: parked vehicles are natural infrastructures. Int. J. Distributed Sens. Netw. **8**(12), 151795 (2012)

5. CarSharing, *The Carsharing Association* [online], 2016 [2018-07-08]. Available on the Internet: http://carsharing.org/

6. Mobypark, *Le parking des voyaguers*. [online], 2016. Mobypark [2018-07-08]. Available on the Internet: https://www.mobypark.com/fr.

7. V. Adamec, et al., Issues of hazardous materials transport and possibilities of safety measures in the concept of smart cities. in: *EAI Endorsed Transactions on Smart Cities*. Italy: EAI, ICST. org. Extended version, 16(1): e4. pp. 1-11 (2016)

8. K. Pradeep, What's the really mean of Smart City?, in: *Smart City Projects* [online], 2015 [2018-06-08]. Available on the Internet: http://www.smartcitiesprojects.com/whats-the-real-mean-of-smart-city/

9. ATEM, *Ateliér ekologických modelů, s.r.o.* [online], 2018 [27 Sep 2018]. Available on the Internet: http://www.atem.cz

10. MOVES, *U.S. Environmental Protection Agency* [online], 2018 [2018-10-02]. Available on the Internet: https://www.epa.gov/moves

11. PTV Group, *The Vision Traffic Suite* [online], 2018 [2018-10-02]. Available on the Internet: http://vision-traffic.ptvgroup.com/en-us/products/

12. Emisia, *Conscious of transport's impact* [online], 2018 [2018-10-04]. Available on the Internet: https://www.emisia.com

13. CDV, *Centrum dopravního výzkumu* [online], 2018 [2018-09-30]. Available on the Internet: https://www.cdv.cz

14. RDE, *The Real Driving Emissions* [online], 2017 [2018-9-27]. Available on the Internet: http://www.caremissionstestingfacts.eu/rde-real-driving-emissions-test/

15. SEKO, *Měření emisí motorových vozidel za jízdy* [online], 2015 [2018-09-30]. Available on the Internet: http://www.sekobrno.cz/vyzkum-a-vyvoj/mereni-emisi-motorovych-vozidel-za-jizdy/ (In Czech)

16. RICARDO, *Real Word Emission Monitoring* [online], 2018. Ricardo PLC [2018-09-10]. Available on the Internet: https://ricardo.com/news-and-media/press-releases/ricardo-launches-real-world-vehicle-emissions-moni

17. G.B. Hua, *Smart Cities as a Solution for Reducing Urban Waste and Pollution* (Information Science Reference, Hershey, 2016). An Imprint of IGI Global. ISBN 978-1-5225-0302-6

18. Metodika Konceptu inteligentních měst, *Project TB930MMR001* [online], 2015 [2018-09-17]. Available on the Internet: http://www.mmr.cz/getmedia/b6b19c98-5b08-48bd-bb99-756194f6531d/TB930MMR001_Metodika-konceptu-Inteligentnich-mest-2015.pdf   (In Czech)

19. Climo, *Bosch Climo* [online], 2017 [2018-09-18]. Available on the Internet: http://www.boschclimo.com

20. M.S. Jamil et al., Smart environment monitoring system by employing wireless sensor networks on vehicles for pollution free smart cities. Procedia Eng. **107**, 480–484 (2015)

21. A. Goel, et al., Air pollution detection based on head selection clustering and average method from wireless sensor network, in: *Advanced Computing & Communication Technologies (ACCT), 2012 Second International Conference* (IEEE, 2012), pp. 434–438. ISBN: 9781467304719

22. A. Miles et al., IoT-based decision support system for monitoring and mitigating atmospheric pollution in smart cities. J. Decision Syst. Taylor & Francis **27**, 56–67 (2018). ISSN: 1246-0125

23. BreezoMeter, *Location-based, real-time, air quality data you can trust* [online], 2018 [2018-09-18]. Available on the Internet: https://breezometer.com

24. Y. Mehta, et al., Cloud enable air quality detection, analysis and prediction: a smart city application for smart health, in: *3rd MEC International Conference on Big Data and Smart City*, 2016, pp. 272–278. doi: https://doi.org/10.1109/ICBDSC.20167460380.

25. R.H. Narashid, W.M.N.W. Mohd, Air quality monitoring using remote sensing and GIS technologies, in: *Proceedings of IEEE International Conference on Science and Social Research*, December (2010), pp. 1186–1191

26. V. Adamec, B. Schüllerová, A. Babinec, D. Herman, J. Pospíšil, Using the DataFromSky System to Monitor Emissions from Traffic. In *Transport Infrastructure and Systems: Proceedings of the AIIT International Congress on Transport Infrastructure and Systems*, Rome, Italy, 10-12 April 2017 (Leiden, Netherlands, 2017). pp. 913–918. ISBN: 9781138030091

27. E. Bochinski, V. Eiselein, T. Sikora, *High-Speed Tracking by Detection Without Using Image Information* [online], 2017. [2018-09-08]. Available on the Internet: http://elvera.nue.tu-berlin.de/files/1517Bochinski2017.pdf

28. D. Espejel-Garcia, J.A. Saniger-Alba, G. Wenglas-Lara, et al., A comparison among manual and automatic calibration methods in VISSIM in an Expressway (Chihuahua, Mexico). Open J. Civil Eng. **7**, 539–552 (2018). ISSN: 2164-3172

29. R. Girshick, *FAST R-CNN* [online], 2015 [2018-10-01]. Available on the Internet: https://www.cv-foundation.org/openaccess/content_iccv_2015/papers/Girshick_Fast_R-CNN_ICCV_2015_paper.pdf

30. Ministerstvo životního prostředí [online]. Praha, 2018 [2018-09-21]. Available on the Internet: https://www.mzp.cz (In Czech)

31. H. Okuda, K. Harada, T. Suzuki et al., Modeling and analysis of acceptability for merging vehicle at highway junction, in: *2016 IEEE 19th International Conference on Intelligent Transportation Systems (ITSC)*, 1–4 Nov 2016, Rio de Janeiro. ISSN: 2153-0017

32. J. Park et al., An application of emission monitoring system based on real-time traffic monitoring. Int. J. Inf. Process. Manag. **4**(1), 51–57 (2013). https://doi.org/10.4156/ijipm.vol4.issue1.7. Gyeongju-si: Advanced Institutes of Convergence Information Technology. ISSN 20934009

33. N. Raju, P. Kumar, A. Chepuri, et al., Calibration of vehicle following models using trajectory data under heterogenous traffic conditions, in *Conference: Transportation Research Board 96th Annual Meeting, Transportation Research Board*, Vol. 17-05479, Washington DC, USA (2017)

34. S. Ren, K. He, R. Girshick, J. Sun, *Faster RCNN: Towards Real-Time Object Detection without Region Proposal Networks* [online], 2012 [2018-09-08]. Available on the Internet: https://papers.nips.cc/paper/5638-faster-r-cnn-towards-real-time-object-detection-with-region-proposal-networks.pdf

35. S.M.S. Mahmud et al., Application of proximal surrogate indicators for safety evaluation: A review of recent developments and research needs. IATSS Res. **41**(4), 153–163 (2017). ISSN: 0386-1112

36. SYMOS'97, *IDEA-ENVI* [online], 2017 [2018-09-27]. Available on the Internet: https://www.idea-envi.cz/symos-97.html

37. D. Sperling, D. Gordon, Two Billion Cars: Transforming a Culture, TR News 259, 3–9 (2008).

38. A. Cathcart-Keays, *Will we ever a realy car-free city?* [online], 2015 [2018-10-01]. Available on the Internet: https://www.theguardian.com/cities/2015/dec/09/car-free-city-oslo-helsinki-copenhagen

39. EURACTIVE. *Air Quality* [online], 2016 [2018-10-01]. Available on the Internet: https://www.euractiv.com/sections/air-pollution/

40. EUROSTAT. *Greenhouse gas emissions from transport* [online], 2018 [2018-10-03]. Available on the Internet: https://www.eea.europa.eu/data-and-maps/indicators/transport-emissions-of-greenhouse-gases/transport-emissions-of-greenhouse-gases-11

41. Data From Sky [online], 2018 [2018-10-03]. Available on the Internet: http://datafromsky.com/platform/software/

Chapter 8
A Combined Data Analytics and Network Science Approach for Smart Real Estate Investment: Towards Affordable Housing

E. Sandeep Kumar and Viswanath Talasila

1 Introduction

Affordable housing is emerging as a major requirement due to the growth in the need for creating equality in the living standards of people in our society [1]. Government is extending the collaboration towards public–private partnership, especially realtors to construct houses and apartments for people who are below the median line of income. The locations for construction of such houses are dependent on numerous real estate attributes, which include social, cultural, economic, physical, and governmental [2]. Some of these attributes include public transportation facilities, availability of public schools and colleges, availability of water and supportive weather conditions, availability of hotels and restaurants, and so on [3]. As the attribute number list grows, so does the complexity of decision-making in location choice. Identification of best locations is an important requirement from the perspective of not only house construction but also purchase of the existing house and renting.

Hence, to understand the trends and solve the above mentioned problems in real estate, many methods and tools are used which are derived from various fields like data science, network science, statistics, probability theory, estimation theory, and so on. Today, the availability of huge volumes of data has paved the path for researchers to use concepts and tools of data analytics to discover goal-oriented knowledge from the existing data. One such important application is hedonic modeling, where the dependency of the various attributes on the real estate price is computed using basic regression methods and machine learning techniques. Use of linear and logistic regression, clustering techniques, support vector machines, and

E. Sandeep Kumar (✉) · V. Talasila
Department of Telecommunication Engineering, M.S Ramaiah Institute of Technology,
Bengaluru, Karnataka, India
e-mail: viswanath.talasila@msrit.edu

© Springer Nature Switzerland AG 2020
N. V. M. Lopes (ed.), *Smart Governance for Cities: Perspectives and Experiences*,
EAI/Springer Innovations in Communication and Computing,
https://doi.org/10.1007/978-3-030-22070-9_8

artificial neural networks [4–14] on the real estate multiple listing service data is often encountered in the literature. Additionally, predicting the house price using data analytic tools is also a well-studied research area. Regression techniques, neural networks, and gradient boosting [15–17] are few tools that are often used. The increasing availability of voluminous data has attracted network science practitioners to infer on the relational status and mutual dependencies among the various data ingredients. Use of social media like Facebook, Twitter, LinkedIn, and various search engines is generating digital footsteps which are used to develop knowledge graphs. These graphs (networks) make us understand the behavior of entities in a network [18–20].

To the best of the authors' knowledge, there has been no focused effort to use network science for the analysis of real estate networks and in specific for location identification. Majority of the software [21, 22] ask the user to enter an exact location, and based on the database query search it will identify suitable condominiums (apartment complexes) for investment. In addition, the existing literature on data analytics in real estate investment is based on the assumptions that a user already knows a location. There are many reasons why an investor may not know the specific location for investment. A simple reason may be that an investor is new to the city. A more involved reason is that even though an investor is native to the city, it is logically impossible to narrow down to a very specific location; at best a small geographical area can be identified. However, in big cities even a small area can easily compromise thousands of dwellings and commercial property; further, even the small area is often highly heterogeneous (in terms of people, establishments, facilities, etc.). Focusing only on price trends does not address the multiple concerns of an investor [15].

Choosing a good location for investment is very crucial since it is dependent on a large number of user's requirements. It may be based on job availability, economic status of people, availability of restaurants, low criminal activities and safety, public transportation facility, availability of schools and shopping malls, and many more. These multiple attributes make a user's decision to select a location more complex and difficult. Under the influence of this huge number of attributes, the location selection may tend towards suboptimal decisions. Hence, an intelligent way of choosing the locations is of greater need in real estate investment that also focuses on the selection of best attributes among that huge number of attributes. Moreover, the mutual influence of the attributes is not considered in the existing literature and attributes are assumed to be independent which is actually not true in general.

In this chapter, a novel algorithm has been presented that selects best attributes for a user and based on user's choice the algorithm identifies the best locations (throughout this chapter, authors refer to condominium complexes as locations) for investment in real estate. In the proposed work, nearly 200 real estate attributes were considered and the best attributes were selected based on the metric called χ which is computed mainly based on the Pearson correlation coefficient [23]. The obtained attributes form a source of choice for a user and based on his/her selections layers of machine learning techniques are activated. In the first layer roads and streets are identified and in layer 2 condominiums (locations) in that street is spotted. For this purpose, statistical modeling with machine learning techniques is used which are taken from data science as analytic tools. However, applications of data analytics succeed well in identification and classification problems; a clearer inference on the relational status of the attributes cannot be obtained, especially when the entities are in huge number. Hence, a bipartite network is added as an extension to the machine learning layers that provides a relational status of the attributes and their influence on the various streets and roads (collectively we call them as landmarks hereafter), while selecting best location among the shortlisted locations by the machine learning layers, for investment. The network uses eigen centrality on a bipartite network of attributes and condominiums for selecting the best condominium for investment. The advantage of using the network layer is the selection of condominium based on the relationship status of all the other attributes with the condominiums in the network. An example simulation on the analysis of network dynamics is provided to leverage the advantages of network science by creating perturbations in the link weights of the network, to find the influential and stable attribute in the designed real estate bipartite network.

The location identification algorithm is tested on the data obtained from the official database called *TerraFly* [24] which is created and maintained by Florida International University (FIU). We have restricted this work to nine landmarks. Rest of this chapter is organized as follows: Sect. 2 deals with the applications of data and network science for smart governance; Sect. 3 discusses about the related works and the state-of-the-art comparison with the current work in the chapter; Sect. 4 deals with the data set used in this work and the assumptions on the work; Sect. 5 discusses the statistical modeling used to identify the best attributes, and discusses the use of decision trees and PCA and K-means clustering for location identification; Sect. 6 deals with the network science for location identification problem; Sect. 7 discusses the obtained results and discussions; Sect. 8 discusses the implications of the obtained result on the smart governance; and finally Sect. 9 discusses the conclusions of the work.

1.1 Scopes of This Work

- Application of statistical modeling to obtain the best attributes from nearly 200 real estate attributes based on the metric χ which is computed using Pearson's correlation coefficient.
- Use stacks of machine learning algorithms to identify locations (condominiums) for investment.
- Application of network science to obtain the best condominium based on centrality measures by constructing a real estate network.
- Simulation study on the most consistent and influential attribute in the presence of link weight perturbations in the designed real estate network.

2   Data and Network Science Applications in Smart Governance

Technology has already made its way into our daily activities. Shopping, transportation, communication, education, health, and so many other things rely on smart devices that are driven by advanced computing and techniques. Usage of such gadgets has paved the way to accumulate a large amount of user data which enabled the era of big data analytics [25]. Analysis of and drawing useful inferences based on various applications can help in the betterment of society. However, operating on such voluminous and versatile data has to be carried out using sophisticated methods. Data science is one such solution enabler [26] which is an interdisciplinary area comprising the tools for statistics, mathematics, probability theory, artificial intelligence, and machine learning to a larger extent and provides us better understandability of the data for various applications and one such application is *smart governance*. Applications in smart governance include consumer behavioral analytics, natural calamity predictions, crime predictions, social service-related analytics, healthcare analytics, data security and privacy, finance, and bank analytics.

Network science is another interdisciplinary research area that draws the theories and methods from graph theory, mathematics, statistics, physics, sociology, and data mining. The current-day computer networks, Internet, and various communication networks are analyzed using network analysis. Similarly, there are other few interesting applications of the network science (complex network analysis) that find their profound applications in smart governance including biological networks [27], transportation networks, epidemic networks, social networks, financial networks, and so on. Readers who are further interested to know about complex networks are directed to read [28]. Even though this chapter highlights the use of data and network science to solve the real estate location identification for housing application, there are still numerous other tasks in smart cities and governance that can be addressed using the able usage of data and network sciences separately or by the combination of both. Two specific examples of smart governance applications are discussed in detail in Sect. 8.

3   Related Works and State-of-the-Art Comparison

The application of data analytics and network science to solve the real estate location identification problem is novel. The existing works in data analytics focus much on the prediction and modeling of the real estate price. Authors in [29] propose a method to predict the house price index (HSI) using data analytic techniques like clustering and principal component analysis for Kookmin bank data. Work in [4] discusses the use of linear regression to find the relation between the real estate price and the attributes; for this purpose authors use real estate data of Harbin city. In [30], authors propose a linear regression hedonic model to find the spatial dependency of the real estate price; for this purpose they have considered the real estate data of eight countries from the multiple listing service database.

In [31], authors discuss a framework of using fuzzy logic to find the selling price of a real estate property in the presence of incomplete information. They claim that the developed method helps in reducing the risk that arises from the uncertainties in the input. In [32], authors use fuzzy logic systems for hedonic house price modeling. In [33], authors use network science to determine the financial activities among different entities. This analysis would help in revealing financial crimes like terrorism, narcotic laundering, and so on.

Authors in [34] discuss the usage of network science to model the supply chain in the form of a network. Authors claim that such kind of structural studies helps in revealing the robustness of supply chains and reveal the topological behavior of the supply chains while overcoming the limitations of the existing supply chain networks. In [35], authors use complex network analysis in considering the micro- and macro-level attributes of revenue of UK stock market and forecast the stock value predictions using complex networks.

There have been limited applications of data science towards real estate location identification problem. The trend majorly includes hedonic price modeling that relates the real estate price with its attributes and prediction of real estate price using the attributes. There are no specific trends in network science applications for real estate investment-related problems.

By carefully traversing through the existing works, as quoted in Sect. 1 and in the current section, it is clear that all the existing works in data analytics focus more on the predictions of the price and hedonic modeling; the location identification for investment has not been explored much. In addition, there are no specific works of network science usage that focus on the issues in the real estate investment. There are few applications in the area of finance but there are no confined works that focus on the real estate. The existing software and companies of real estate investment [36–38] either focus on the construction of building, maintaining, and looking after the documentation works or suggest the investment locations based on some database queries. All these methods in the existing works do not apply any machine learning techniques that are based on the query approach. Hence, considering the real estate investment location identification problem the work proposed in this chapter is a novel attempt that combines data and network science under a single roof to find a robust solution for the given problem of location identification for affordable housing in real estate investment.

4 Data Set

The data is obtained from TerraFly database [24] managed and maintained by Florida International University (FIU) in collaboration with the US Government. The database is a big data platform and a query-based system with complete information regarding economic, social, physical, and governmental factors of selected countries.

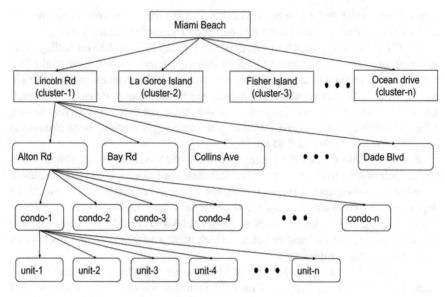

Fig. 8.1 Hierarchical view of the available data (clustered)

The landmarks (we call streets, roads, and so on as landmarks in this work) of Miami Beach are divided into clusters. Preference is given to nearby landmarks while clustering, however, can also be random. For our initial work, single cluster with nine landmarks and their associated real estate data (available as multiple listing service (MLS) data) is considered. They are condominium (also called as condos) data of Alton Rd, Bay Rd, Collins Ave, Dade Blvd, James Ave, Lincoln Rd, Lincoln CT, Washington Ave, and West Ave. These landmarks belong to Miami Beach City of Miami Dade County, FL, USA. The approximate count of condominiums was obtained from the official database of Miami Beach, i.e., for Alton Rd-7000 condominiums, Bay Rd-7000, Collins Ave-9000, Dade Blvd-1500, James Ave-2000, Lincoln Rd-2000, Lincoln CT-2000, Washington Ave-4000, and West Ave-2000, respectively. The hierarchical view of the data is shown in Fig. 8.1. For our analysis from every landmark, 500 condominium data were randomly picked for training and 500 for validation. The processes of training and validation were repeated in five sets and the average validation accuracy is quoted, which will be discussed in detail in the results section.

4.1 Assumptions in This Work

It is assumed that a user is not fully aware of the city location details. He/she has a very little idea about the locations, but do not know whether it is best or not for investment. In addition, a user has assumed a set of attributes; however, they need not be optimal. However, a user should at least know which cluster of landmarks should be opted for his/her investment.

5 Identification of Location Using Data Analytics

In this section, a detailed analogy of the attribute selection, stacked layers of machine learning for location identification, and real estate network with its related analytics is discussed.

5.1 Attribute Selection

Real estate comprises a large list of attributes that can be broadly classified into economic, social, physical, and governmental [24]. The current framework is bound to the real estate attributes; however, the same method can be extended for other factors as well.

Out of a large number of attributes, the best set of attributes are selected based on the value χ, which is a representative of the strength of an attribute in a landmark in real estate investment. The calculation of χ is as follows:

$$\chi = w_1 C + w_2 N \tag{8.1}$$

where w_1 and w_2 are constants and called the weights, C is the Pearson coefficient of an attribute with the real estate price, and N is the number of data sample points available in an attribute after cleansing. Here, χ is an identity number assigned to every attribute in a landmark, based on which a top attribute is selected. For this calculation, the Pearson coefficient is used as an initial choice. However, there are other correlation metrics such as Spearman and Kendall coefficients [39] that can also be used instead. The selection algorithm is as follows:

Algorithm 1

1. Start
2. Collect the condominium data of all landmarks in a cluster. Initialize w_1 and w_2.
3. For first landmark, find the parameter χ which determines the relation between the first attribute of a condominium and real estate price.
4. Repeat the experiment for all the attributes. Select the top k number attributes from every condominium. Select the top u number of attributes based on the number of occurrences in a landmark.
5. Repeat steps 2 and 3 for all landmarks.
6. Combine all u and select top y attributes based on number of occurrences. This set is the optimal attribute set for that cluster of landmarks.
7. Repeat this process for all clusters of landmarks.
8. End.

For simulations, the cluster of nine landmarks was chosen and the parameters were set to $k = 10$ and $u = 10$, i.e., choosing top ten attributes every time. However, $y = 9$ is the count of the number of attributes for the entire cluster. The simulation is repeated for five iterations with each time 500 condominiums selected in random from every landmark (training data set). The top nine attributes obtained are as follows:

- **Number of beds**: Number of bedrooms available in the unit of a condominium building.
- **Number of full baths:** Number of full bathrooms (tub, shower, sink, and toilet) available in the unit.
- **Living area in sq. ft.:** The space of the property where people are living.
- **Number of garage spaces:** Number of spaces available for parking vehicles.
- **List price:** Selling price of the property (land + assets) to the public.
- **Application fee:** Fee paid for owners' associations.
- **Year built:** Year in which the condominium/apartment complex is built.
- **Family limited property total value 1:** The property value accounted for taxation after all exemptions. This is for the district that does not contain schools and other facilities.
- **Tax amount:** The amount paid as tax for the property every year.
- When a user chooses this cluster of nine landmarks, the above attributes will be given to him/her as choices. A user can select and set the magnitudes to the attributes according to his/her wish. It is not mandatory that all attributes need to be filled. These attributes are passed onto two layers of machine learning: in the first layer decision trees and in the second layer principal component analysis with K-means clustering. In addition, the rationale behind the choice of a multilayer classification model is provided in [40]; interested readers are suggested to refer to that article.

5.2   Decision Trees in Layer 1

In this section, the use of decision trees [41, 42] to select the best landmark for investment is dealt with detail. The working of tree follows the naive ID3 algorithm [43].

The attribute set used in a tree may change depending upon the landmarks in that cluster selected. The top nine attributes of any cluster have χ values in every landmark obtained by averaging the χ values of all the condominiums in that landmark. These χ values are compared with the other landmarks and the highest χ value is retained with its respective landmark. The result in Table 8.1 is the output of Algorithm 1 on the training set (500 condominiums selected randomly from every landmark). The process was repeated for five iterations and in all iterations the landmarks and the attribute pair remained same likewise shown in Table 8.1. The χ values shown are the average of the five iterations. Same results were obtained for the validation set as well. This table serves as a backbone for the decision tree operation.

Table 8.1 χ values of attributes input to decision tree

Attribute	χ value	Landmark
Number of beds	1.338	Alton Rd
Number of full baths	1.380	Alton Rd
Year built	1.226	Lincoln CT
Application fee	1.235	James Ave
Number of garage spaces	1.233	Alton Rd
List price	1.894	James Ave
FLP total value	1.291	Washington Ave
Living area	1.375	Alton Rd
Tax amount	1.164	Bay Rd

A user selects a particular cluster followed by choosing attributes of his/her interest and setting suitable magnitudes for them. For a decision tree, only a user interest vector is used. For example, suppose a user is interested in the number of beds and number of full baths in the same order as in Sect. 5.1; then the user vector is 110000000. This is fed into the decision tree.

An example decision tree with three attributes which is symmetric and binary is as shown in Fig. 8.2. In the figure, every time when the tree traversal hops from one node to another, it considers the χ value (if it is "YES" case) of the current attribute and compares with the χ value of the previous and selects the landmark with the highest χ value. The process continues until one landmark is retained at the end. This technique is called the ***highest magnitude win*** approach. The obtained truth table is shown in Table 8.2.

Notice the last column in Table 8.2. The rows are the different cases (shown are just four cases) in which a user can enter attributes. The generated user vector is the row. For the case "011," which is the third row in Table 8.2, it means a user is interested in number of full baths and list price. The magnitude of χ for these attributes is 1.380 and 1.894, respectively. According to the highest magnitude win strategy, James Ave wins in this case, which is placed in the output landmark column in the corresponding row. This is the leaf of that branch in the tree. The tree and its traversal are shown in Fig. 8.2.

Even though the tree output does not change, the position of the tree node (order of the columns in Table 8.2) matters for the fact that the time taken for the tree to converge to a landmark depends on the node information richness. To identify the positions of the nodes, the ID3 algorithm is adapted and in turn helps us to choose an optimal root attribute. For a truth table that is binary in nature and having all possibilities in it, all attributes have the same information gain according to ID3 and any attribute can be a root. However, if the truth table has chosen possibilities of truth and false, ID3 will help to choose the best root attribute. A detailed procedure on the usage of ID3 for root node selection is provided in APPENDIX-B of [40]. In addition, a decision tree need not be always as shown in Fig. 8.2, where the children nodes are identical, but they can differ as well, based on the landmarks and the cluster considered.

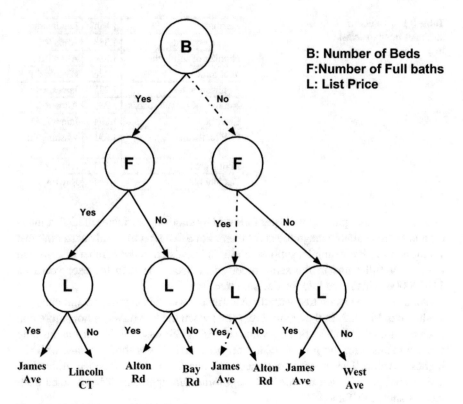

B: Number of Beds
F:Number of Full baths
L: List Price

Fig. 8.2 Decision tree and its traversal

Table 8.2 Truth table for decision tree operation

Number of beds	Number of full baths	List price	Output landmark
0	0	1	James Ave
0	1	0	Alton Rd
0	1	1	James Ave
1	1	1	James Ave

5.3 Principal Component Analysis (PCA) and K-Means Clustering in Layer 2

The previous section discussed the use of decision trees to identify the best landmark among the chosen cluster of landmarks. In this section, layer 2 will be analyzed in detail, which helps to find locations in the landmark, output by layer 1. In the previous section, only the user's interest vector was considered, but in this layer the entered magnitudes are considered for location identification. The steps adapted in the layer 2 are discussed further in detail.

5.3.1   Finding Principal Components

Every landmark has condominiums which have the attributes mentioned as per Sect. 5 (in our study there are nine top attributes). Let us consider the algorithm below:

Algorithm 2

1. Start.
2. Consider first landmark from a cluster.
3. For every condominium in that landmark, find the principal components for the top attributes.
4. Consider the first principal component among the available components and average all the first components over a landmark.
5. Repeat steps 2 and 3 for all landmarks.
6. End.

Let the principal components [44] of a landmark be PC. Using PC of a particular landmark, principal scores of units in that condominium are calculated using (8.2). Every top attribute in that condominium is multiplied by PC:

$$PCs = \sum_{i=1}^{y} attribute\,(i) * PC\,(i). \tag{8.2}$$

where y is the count of top attributes of a cluster of landmarks. On averaging PC_S of all the condominium units, a PC score for a condominium is obtained. Repeat the process for all the landmarks and their associated condominiums. Averaging PC scores of units in a condominium will result in PC score for a condominium. Once the PC scores are available, K-means clustering [45] is applied to the scores of the condominiums of a landmark dividing the condominiums into K groups each having its own centroid.

Layer 2 mainly operates on the magnitude entered by a user which was not considered in layer 1. Since the landmark is already chosen by layer 1, the average PC of that landmark is known. Hence, the user-entered magnitude and the average PC are multiplied to get a PC score using (8.2). This score is compared with the centroids of the groups created by K-means clustering and the group with shortest Euclidean distance is selected as the best condominium group and its ingredient condominiums are rated as the best for a user according to his/her entered choices.

6   Network Science Approach for Location Identification

Data analytics uses systematic ways of modeling and learning the data, which is the digital trace in the ongoing world. However, to understand the network prospects of the data with their mutual influences, combining data with a network is very essential [46]. This interdisciplinary setup makes complex systems like real estate investment more understandable.

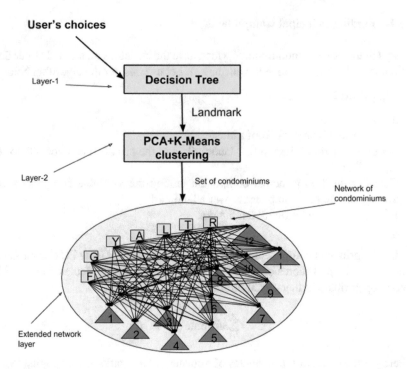

Fig. 8.3 Extended network layer

In this context, a network structure is rigged for the condominiums obtained from the layer 2 machine learning technique. This network view helps a user to decide the best condominium in the presence of mutual relationships among other condominiums with their attributes. The system model is shown in Fig. 8.3.

In Fig. 8.3, a network of condominiums is constructed. These condominiums were the list of condominiums output from layer 2 that comprised of PCA and K-means clustering. The obtained graph is a bipartite network which comprises two parties, viz. attributes and condominiums. It is to be observed that there is no link between the same party members and the link always flows from one party to the other. The attributes are shown in a square shape and the condominiums are shown in a triangular shape. On applying various centrality measures like eigen centrality [47], alpha centrality, closeness, and so on [48], it is able to draw various conclusions on the important condominiums in the network. In this work, eigen centrality is used to obtain the most influential condominium. In terms of real estate, eigenvalue depicts the amplification factor for the influence and the eigenvector infers the direction of the influence in the network.

Addition of network at the end of the machine learning layers is just a demonstration; however, layer 1 or layer 2 can also be visualized as a network, where suitable centrality measures will infer for landmark and location selection in that particular layer. As an example, in layer 1 eigen centrality can be used to select a landmark and in layer 2 alpha centrality to select the set of condominiums.

7   Results and Discussions

This section briefs the obtained results of the proposed methodology and its related discussions in detail.

The attribute selection (Algorithm 1) discussed in Sect. 5.1 was executed on five training and validation data sets. In each case, for the entire nine landmarks, unique attributes were listed. The attribute set was compared between the training and validation set and the number of mismatches was accounted. The obtained results for five datasets selected in random (i.e., five iterations) are available in Table. 8.3. It was observed that the system remains consistent with the training and validation result matching with an average of 96.86% accuracy.

The same process is repeated for the decision trees of layer 1. In the case of the tree, as it is explained already in Sect. 5.2, the tree traverses from one node to another considering the value of χ of every attribute. Hence, the attribute and landmark corresponding to the highest value of χ play a major role in decision tree outputting a landmark. The top attributes as per Sect. 5.1 are listed with its χ value averaged over a landmark in Table 8.4 for five iterations (both training and validation). It is observed that the winning landmarks with highest χ for an attribute remain the same in training and validation, which is the accuracy of 100% and in turn defines the accuracy of the decision tree.

From Table 8.4, the attribute with the landmark having highest χ value remains consistent throughout five iterations (as per both training and validation data sets), with 100% validation accuracy. The highest χ value is highlighted in bold in the table.

In layer 2, principal component analysis and K-means clustering were used to find the best condominiums. These techniques were applied to the training and the validation data set. The centroids of the groups of condominiums obtained after clustering were compared in both training and validation over five iteration data sets. The deviation error (mean absolute error) was noted in each case and the obtained results are available in Table 8.5.

The average validation accuracy of layer 2 from Table 8.5 is 90.25%.

The obtained condominiums from layer 2 were used to construct a bipartite network with attributes and the condominiums as the two parties and χ values linking them. From Table 8.4, it is evident that every attribute has χ value linking landmarks. Hence, a complex network of the attributes and condominiums is constructed

Table 8.3 Accuracy of best attributes selected

Iteration	No. of mismatches	Accuracy
1	1 out of 25	96%
2	0 out of 24	100%
3	1 out of 25	96%
4	1 out of 26	96.15%
5	1 out of 26	96.15%
Average		**96.86%**

Table 8.4 Attributes and their average χ values in every landmark

Iteration	Attributes	Alton Rd	Bay Rd	Collins Ave	Dade blvd	James Ave	Lincoln Rd	Lincoln CT	Washington Ave	West Ave
1 (Training)	Number of beds	**1.34**	1.29	1.22	1.22	1.20	1.20	1.20	1.16	1.24
	Number of full baths	**1.38**	1.30	1.28	1.25	1.14	1.21	1.20	1.27	1.27
	Year built	1.07	1.14	1.17	1.16	1.07	1.21	**1.23**	1.18	1.20
	Application fee	0.75	0.88	0.85	0.75	1.22	0.99	0.97	0.72	0.88
	Number of garage spaces	**1.24**	1.17	1.12	1.07	1.09	1.10	1.11	1.03	1.19
	List price	1.80	1.80	1.73	1.69	**1.89**	1.71	1.73	1.72	1.78
	FLP total value	1.28	1.27	1.24	1.09	0.99	1.09	1.17	**1.32**	1.26
	Living area	**1.37**	1.34	1.26	1.17	1.19	1.23	1.24	1.16	1.29
	Tax amount	1.09	**1.16**	0.93	0.99	0.12	0.84	0.88	1.08	0.99
1 (Validation)	Number of beds	**1.34**	1.30	1.21	1.22	1.19	1.20	1.20	1.16	1.21
	Number of full baths	**1.38**	1.31	1.27	1.26	1.13	1.20	1.20	1.27	1.27
	Year built	1.07	1.12	1.16	1.15	1.05	1.20	**1.23**	1.17	1.20
	Application fee	0.78	0.88	0.85	0.76	**1.24**	0.96	0.97	0.73	0.89
	Number of garage spaces	**1.24**	1.18	1.10	1.06	1.08	1.09	1.13	1.04	1.20
	List price	1.79	1.81	1.72	1.69	**1.89**	1.71	1.72	1.72	1.78
	FLP total value	1.27	1.28	1.23	1.11	0.98	1.08	1.17	**1.32**	1.27
	Living area	**1.38**	1.35	1.27	1.16	1.18	1.23	1.25	1.17	1.29
	Tax amount	1.08	**1.17**	0.88	1.01	0.08	0.82	0.89	1.08	1.00

2 (Training)	Number of beds	**1.34**	1.31	1.22	1.22	1.19	1.21	1.20	1.17	1.24
	Number of full baths	**1.38**	1.32	1.27	1.26	1.14	1.21	1.20	1.28	1.27
	Year built	1.07	1.11	1.18	1.16	1.06	1.2181	**1.2182**	1.19	1.20
	Application fee	0.76	0.86	0.82	0.75	**1.23**	0.95	0.97	0.72	0.90
	Number of garage spaces	**1.23**	1.16	1.11	1.06	1.08	1.10	1.11	1.03	1.20
	List price	1.80	1.80	1.73	1.69	**1.89**	1.71	1.71	1.73	1.78
	FLP total value	1.28	1.28	1.23	1.11	0.98	1.10	1.15	**1.34**	1.27
	Living area	**1.37**	1.35	1.26	1.17	1.18	1.24	1.24	1.18	1.29
	Tax amount	1.09	**1.15**	0.90	1.01	0.11	0.86	0.83	1.12	0.98
2 (Validation)	Number of beds	**1.33**	1.31	1.21	1.22	1.20	1.20	1.20	1.17	1.24
	Number of full baths	**1.37**	1.32	1.27	1.25	1.14	1.20	1.20	1.27	1.26
	Year built	1.08	1.11	1.19	1.15	1.07	1.20	**1.24**	1.18	1.21
	Application fee	0.76	0.87	0.83	0.75	**1.23**	0.93	0.97	0.72	0.89
	Number of garage spaces	**1.23**	1.16	1.11	1.06	1.09	1.10	1.12	1.04	1.18
	List price	1.80	1.80	1.72	1.69	**1.89**	1.70	1.73	1.72	1.78
	FLP total value	1.26	1.28	1.23	1.09	0.98	1.09	1.19	**1.32**	1.26
	Living area	**1.36**	1.35	1.27	1.16	1.19	1.23	1.25	1.17	1.28
	Tax amount	1.06	**1.15**	0.88	0.98	0.11	0.87	0.93	1.07	0.97
3 (Training)	Number of beds	**1.33**	1.30	1.06	1.22	1.19	1.21	1.20	1.17	1.24
	Number of full baths	**1.37**	1.31	1.11	1.25	1.14	1.21	1.20	1.27	1.26
	Year built	1.08	1.13	1.01	1.15	1.07	1.20	**1.23**	1.18	1.21
	Application fee	0.75	0.87	0.76	0.75	**1.24**	0.95	0.96	0.73	0.88
	Number of garage spaces	**1.22**	1.16	0.95	1.06	1.08	1.11	1.11	1.03	1.19
	List price	1.80	1.80	1.50	1.67	**1.89**	1.71	1.72	1.72	1.78
	FLP total value	1.25	1.28	1.07	1.09	0.97	1.09	1.17	**1.32**	1.26
	Living area	**1.36**	1.35	1.09	1.16	1.18	1.23	1.24	1.17	1.29
	Tax amount	1.07	**1.16**	0.77	0.99	0.06	0.82	0.90	1.07	0.97

(continued)

Table 8.4 (continued)

Iteration	Attributes	Alton Rd	Bay Rd	Collins Ave	Dade blvd	James Ave	Lincoln Rd	Lincoln CT	Washington Ave	West Ave
3 (Validation)	Number of beds	**1.34**	1.30	1.20	1.22	1.19	1.20	1.20	1.17	1.25
	Number of full baths	**1.38**	1.31	1.26	1.25	1.13	1.21	1.20	1.28	1.27
	Year built	1.07	1.12	1.16	1.15	1.06	1.20	**1.21**	1.18	1.20
	Application fee	0.78	0.87	0.85	0.76	**1.23**	0.94	0.97	0.72	0.89
	Number of garage spaces	**1.23**	1.17	1.10	1.07	1.08	1.11	1.13	1.04	1.21
	List price	1.80	1.80	1.72	1.69	**1.89**	1.69	1.71	1.73	1.79
	FLP total value	1.28	1.28	1.21	1.10	0.98	1.08	1.15	**1.32**	1.27
	Living area	**1.37**	1.35	1.25	1.16	1.18	1.23	1.24	1.17	1.31
	Tax amount	1.08	**1.16**	0.87	1.00	0.09	0.82	0.85	1.09	0.99
4 (Training)	Number of beds	**1.33**	1.30	0.99	1.22	1.20	1.20	1.20	1.17	1.24
	Number of full baths	**1.37**	1.31	1.03	1.25	1.15	1.20	1.20	1.27	1.26
	Year built	1.07	1.12	0.95	1.15	1.07	1.20	**1.23**	1.17	1.21
	Application fee	0.77	0.86	0.69	0.76	**1.21**	0.96	0.96	0.73	0.90
	Number of garage spaces	**1.23**	1.16	0.91	1.07	1.09	1.10	1.11	1.03	1.18
	List price	1.80	1.80	1.41	1.68	**1.89**	1.72	1.72	1.72	1.79
	FLP total value	1.26	1.28	1.01	1.08	0.99	1.09	1.17	**1.32**	1.27
	Living area	**1.37**	1.35	1.03	1.16	1.19	1.23	1.24	1.18	1.28
	Tax amount	1.08	**1.16**	0.72	0.98	0.15	0.83	0.90	1.07	0.95
4 (Validation)	Number of beds	**1.34**	1.30	1.22	1.22	1.19	1.20	1.20	1.16	1.24
	Number of full baths	**1.38**	1.31	1.27	1.25	1.13	1.21	1.20	1.26	1.27
	Year built	1.07	1.12	1.17	1.15	1.06	1.20	**1.21**	1.17	1.19
	Application fee	0.76	0.87	0.85	0.75	**1.24**	0.93	0.97	0.72	0.90
	Number of garage spaces	**1.24**	1.16	1.11	1.07	1.07	1.11	1.13	1.04	1.20
	List price	1.80	1.80	1.74	1.68	**1.89**	1.70	1.71	1.71	1.79
	FLP total value	1.27	1.27	1.23	1.09	0.97	1.08	1.15	**1.29**	1.26
	Living area	**1.38**	1.35	1.27	1.17	1.17	1.23	1.24	1.16	1.30
	Tax amount	1.08	**1.15**	0.90	0.99	0.06	0.82	0.85	1.04	0.98

5 (Training)									
Number of beds	**1.34**	1.30	1.20	1.22	1.19	1.21	1.21	1.17	1.25
Number of full baths	**1.38**	1.31	1.26	1.26	1.13	1.21	1.21	1.28	1.27
Year built	1.07	1.12	1.16	1.16	1.06	1.20	**1.22**	1.18	1.20
Application fee	0.76	0.88	0.84	0.76	**1.24**	0.95	0.97	0.72	0.88
Number of garage spaces	**1.23**	1.18	1.10	1.07	1.07	1.10	1.12	1.04	1.20
List price	1.79	1.80	1.72	1.69	**1.89**	1.71	1.71	1.72	1.77
FLP total value	1.26	1.28	1.22	1.10	0.97	1.09	1.15	**1.32**	1.27
Living area	**1.37**	1.35	1.25	1.16	1.17	1.23	1.24	1.17	1.30
Tax amount	1.09	**1.16**	0.88	1.00	0.06	0.84	0.85	1.09	1.00
5 (Validation)									
Number of beds	**1.34**	1.30	1.20	1.22	1.20	1.20	1.20	1.16	1.23
Number of full baths	**1.38**	1.31	1.26	1.25	1.14	1.21	**1.24**	1.27	1.26
Year built	1.07	1.13	1.15	1.15	1.07	1.203	1.22	1.18	1.22
Application fee	0.76	0.87	0.85	0.75	**1.23**	0.946	0.966	0.72	0.90
Number of garage spaces	**1.23**	1.18	1.07	1.07	1.09	1.11	1.13	1.20	1.19
List price	1.80	1.80	1.71	1.68	**1.89**	1.70	1.72	1.73	1.78
FLP total value	1.27	1.28	1.20	1.09	0.98	1.07	1.19	**1.34**	1.26
Living area	**1.38**	1.35	1.23	1.17	1.19	1.23	1.25	1.17	1.28
Tax amount	1.09	**1.16**	0.87	1.00	0.10	0.82	0.94	1.10	0.97

Table 8.5 Deviation error of centroids

Iteration	Alton Rd	Bay Rd	Collins Ave	Dade Blvd	James Ave	Lincoln Rd	Lincoln CT	Washington Ave	West Ave
1	13.11%	16.88%	7.9%	10.6%	5.8%	11.3%	6.4%	4.6%	17.0%
2	11.46%	12.72%	14.1%	11.0%	1.9%	7.1%	6.6%	6.7%	8.1%
3	10.12%	7.79%	10.0%	11.3%	18.3%	7.4%	10.7%	5.1%	7.3%
4	11.02%	7.69%	1.9%	7.1%	10.4%	12.0%	11.5%	5.3%	15.3%
5	5.215%	11.60%	6.9%	7.7%	10.3%	11.5%	10.0%	3.9%	26.1%
Avg. error	10.18%	11.3%	8.2%	9.6%	9.3%	9.9%	9.0%	5.1%	14.8%
Correct clustering	89.8%	88.6%	91.7%	90.3%	90.6%	90.1%	90.9%	94.8%	85.2%

and eigen centrality is applied to check the best condominium.Let us consider following values for the attributes: number of garage spaces = 3, application fee = 400, number of full bathrooms = 3, number of bedrooms = 2, built year = 1986, taxable property value = 1,942,446, living area = 1007 Sq. ft, tax amount = 8633, and list price = 2,000,000. It was observed that the landmark selected by layer 1 was James Ave, and 401 condominiums were selected by layer 2 in that landmark. After applying eigen centrality, out of 401 condominiums condominium-1701 was selected as the most central condominium. The obtained complex network of real estate scenario including attributes and landmarks is shown in Fig. 8.4, in which the blue-colored circles are the attributes and the red-colored triangles are the condominiums. The links are having varying colors based on their weights (χ magnitude). If the link weight is more than 1 then the color is green; else it is blue.

One of the important purposes of using network science is to study the relationship between the condominiums and the attributes, and their influences on each other. Consider another example, where the attributes are set for simulations like the following: number of beds = 2; number of garage spaces = 2; number of full bathrooms = 2; and application fee = 126. According to the proposed algorithm, Alton Rd was the result of the decision tree and 169 condominiums were selected by layer 2. According to eigen centrality condominium-6487 was selected as the best condominium. The obtained network is as shown in Fig. 8.5.

To know which attribute remains consistent under the link weight variations, the link weights of the network in the Fig. 8.5 are varied (such that percent of the link weight was added to the weight as random noise) and each time the eigen centrality values were noted. At 0% variation (or wihtout variations) of link weights, the obtained centrality values are as shown in Table 8.6. The link weights were added in steps of 10% of their existing weights and the changes in the centrality measures of the network nodes were observed.

It was observed that as the link weight increases, F's centrality value also increases and at a point of 30% increase L loses its central position and F becomes the more central attribute (Fig. 8.6). Hence, it was concluded that adjusting the weight of the links controls the centrality of the nodes in the network. This also implies that the more the correlation of an attribute with the real estate price of a condominium in a landmark, the more that attribute will become central in the

Fig. 8.4 Complex real estate network of condominiums of James Ave

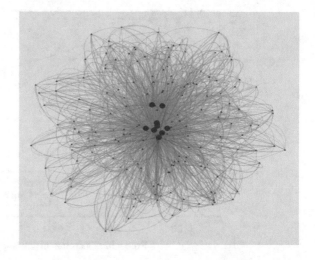

Fig. 8.5 Complex real estate network of condominiums of Alton Rd

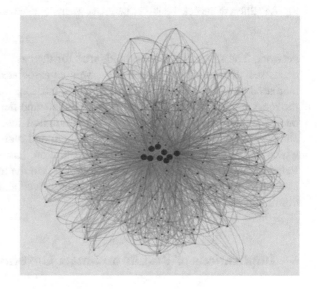

Table 8.6 Eigen centrality values at 0% change in the link weight

Attribute	Eigen centrality value
Year built (Y)	0.4426
Number of garage spaces (G)	0.5621
Tax amount (X)	0.6645
Application fee (A)	0.6921
Living area (R)	0.7502
FLP total value 1 (T)	0.7648
Number of bedrooms (B)	7651
Number of full baths (F)	0.7787
List price (L)	1.0000

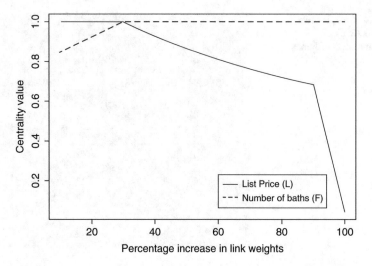

Fig. 8.6 Effect of the link weight change on the centrality values of the top two attributes

network. The same explanation holds true for the case when the link weights are decreased. In another experiment, the weights associated with all attributes were increased to a maximum 10% of their values' existing weights randomly, to check the most stable attribute. This helps us to understand the most consistent attribute due to sudden uncertain inflations. This simulation indicated that during sudden changes in the correlation between an attribute and real estate price, i.e., χ value, which may be due to natural calamities, inflation, and so on, list price attribute remains stable attribute by being most influential in the real estate investment. The same analogy can be drawn on condominiums as well and the most consistent condominium can be found.

8   Implications of Results on Smart Governance

In this section, two major applications of the results obtained due to this research study are highlighted. Firstly, the impact of the various factors on major public utility centers in a city (Fig. 8.7): The public utilities may be bus stations, airports, train stations, shopping malls, highways, and harbors and the factors include weather, ongoing protest/strikes, and road traffic. It is clear that in this network, there are two parties: the public utilities and the factors influencing them. Hence, a bipartite network will give a relationship between these parties. The thickness of the edges indicates the intensity of impact on the entity by that respective factor. From the sample network it is clear that the shopping malls in the city are highly impacted by all these factors and the harbor is least affected. A detailed analysis of how the various shopping malls are affected by the factors will give a detailed analysis of the

Fig. 8.7 Impact network (*W* weather, *A* airport, *T* train station, *B* bus station, *P* ongoing protest, *H* harbor, *S* shopping mall, *Y* highway, *R* road traffic)

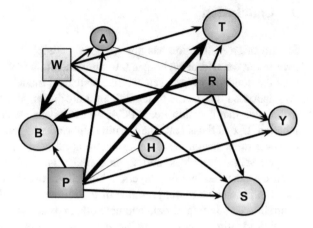

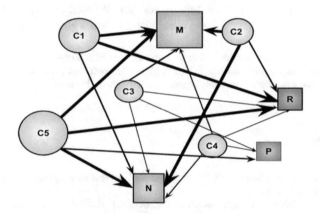

Fig. 8.8 Crime network (*N* molestation, *M* murder, *R* rape, *P* robbery)

relationship. These kind of inferences can be drawn in the same lines as that of the real estate network analysis. All factors influencing the entities may be considered or just top factors/attributes depending on the requirements of the study. If there are many attributes influencing the entities, the best attributes can be selected using the technique mentioned in the above sections.

Another interesting application is the crime studies. In Fig. 8.8, there are five criminals namely C1–C5 and the criminal activities in which they were indulged include murder, robbery, rape, and molestation. The network studies on the graph inference that murder is the most central activity performed by these criminals in that city and among all C5 turns out to be central. Knowing the inference we can go a step deeper and analyze the murder activity in the city in more detail. These are two among numerous scenarios that could be studied using a combination of data and network science.

9   Conclusion

Identification of location has always been a complex task for a user in real estate investment for affordable housing. In this context, the work proposed in this chapter uses the concepts of data science like statistical modeling which uses the Pearson correlation as the means to identify best attributes, and stacked machine learning techniques like decision trees, PCA, and K-means clustering for identification of location. Use of these machine learning algorithms is just a demonstration use case; however other techniques like artificial neural networks, deep learning networks, support vector machines, and so on can also be used. For the locations obtained from the machine learning layers, a bipartite network is constructed and the best location is selected in the presence of the influence of other attributes using eigen centrality. Combining of data and network analytics to obtain more insight into the location identification problem has not been explored much in the existing literature. Hence, this combination provides a more comprehensive approach to affordable housing in real estate investment. In addition, the methodology and the results obtained can be adapted for solving other issues in smart governance.

Acknowledgements Authors would like to thank Dr. Naphtali Rishe and Dr. S.S. Iyengar of School of Computing and Information Sciences, Florida International University, Miami, Florida, for providing the database and valuable suggestions throughout this work.

References

1. Internet document—"Affordable Housing in India", An Inclusive Approach to Sheltering the Bottom of the Pyramid
2. J.F. Schram, *Real Estate Appraisal* (Rockwell, Bellevue, 2006)
3. D.H. Carr, J. Lawson, J. Schultz, Dearborn Real Estate Education, *Mastering Estate Appraisal* (Dearborn Financial Publications, Chicago, 2003)
4. Y. Zhang, S. Liu, S. He, Z. Fang, Forecasting research on real-estate prices in Shanghai, in: *2009 International Conference on Grey Systems and Intelligent Services (GSIS 2009)*, Nanjing, 2009, pp. 625-629
5. W. Wei, T. Guang-ji, Z. Hong-rui, Empirical analysis on the housing price in Harbin City based on hedonic model, in: *2010 International conference on Management Science and Engineering 17th Annual Conference Proceeding*, Melbourne, VIC, 2010, pp. 1659–1664
6. B. park, J.K. Bae, Using machine learning algorithms for housing price prediction: "The case of Fairfax County", Virginia housing data. Expert Syst. Appl. **42**(6), 2928–2934 (2015). ISSN: 0957-4174
7. H. Xue, The prediction on residential real estate price based on BPNN, in: *2015 8th International Conference on Intelligent Computation Technology and Automation (ICICTA)*, Nanchang, 2015, pp. 1008–1013
8. B. Liu, B. Mavrin, D. Niu, L. Kong, House price modeling over heterogeneous regions with hierarchical spatial functional analysis, in: *2016 IEEE 16th International Conference on Data Mining (ICDM)*, Barcelona, 2016, pp. 1047–1052
9. C. Cheng, X. Cheng, M. Yuan, K. Chao, S. Zhou, J. Gao, L. Xu, T. Zhang, A novel architecture and machine learning algorithm for real estate. Signal Inf. Process. Netw. Comput. **473**, 491–499 (2017). Springer, Singapore. Lecture Notes in Electrical Engineering

10. Kecheng Zhao, Wei Shen, Spatial characteristic with individual house properties and multi-level approach to hedonic models, in: *2011 International Conference on Computer Science and Service System (CSSS)*, Nanjing, 2011, pp. 2579–2582
11. T. Oladunni, S. Sharma, Hedonic housing theory—a machine learning investigation, in: *2016 15th IEEE International Conference on Machine Learning and Applications (ICMLA)*, Anaheim, CA, 2016, pp. 522–527. doi: https://doi.org/10.1109/ICMLA.2016.0092
12. B. Park, J.K. Bae, Using machine learning algorithms for housing price prediction. Expert Syst. Appl. **42**, 2928–2934 (2015)
13. I.D. Wilson, S.D. Paris, J.A. Ware, D.H. Jenkins, Residential property price time series forecasting with neural networks, in: *The Twenty-First SGES International Conference on Knowledge Based Systems and Applied Artificial Intelligence*, Cambridge, December 2001, pp. 17–28, Springer Publications
14. H. Xu, A. Gade, Smart real estate assessments using structured deep neural networks, in: *2017 IEEE SmartWorld, Ubiquitous Intelligence & Computing, Advanced & Trusted Computed, Scalable Computing & Communications, Cloud & Big Data Computing, Internet of People and Smart City Innovation (SmartWorld/SCALCOM/UIC/ATC/CBDCom/IOP/SCI)*, San Francisco, CA, 2017, pp. 1-7
15. S. Lu, Z. Li, Z. Qin, X. Yang, R.S.M. Goh, A hybrid regression technique for house prices pre-diction, in: *2017 IEEE International Conference on Industrial Engineering and Engineering Management (IEEM)*, Singapore, 2017, pp. 319–323
16. D. Sangani, K. Erickson, M. A. Hasan, Predicting zillow estimation error using linear regres-sion and gradient boosting, in: *2017 IEEE 14th International Conference on Mobile Ad Hoc and Sensor Systems (MASS)*, Orlando, FL, 2017, pp. 530–534
17. W.T. Lim, L. Wang, Y. Wang, Q. Chang, Housing price prediction using neural networks, in: *2016 12th International Conference on Natural Computation, Fuzzy Systems and Knowledge Discovery (ICNC-FSKD)*, Changsha, 2016, pp. 518–522
18. J. Demongeot, H. Pempelfort, J.M. Martinez, R. Vallejos, M. Barria, C. Taramasco, Information design of biological networks: application to genetic, immunologic, metabolic and social net-works, in: *2013 27th International Conference on Advanced Information Networking and Applications Workshops*, Barcelona, 2013, pp. 1533–1540
19. D.P. Cheung, M.H. Gunes, A complex network analysis of the United States air transportation, in: *2012 IEEE/ACM International Conference on Advances in Social Networks Analysis and Mining*, Istanbul, 2012, pp. 699–701
20. Q. Xuan, Z.Y. Zhang, C. Fu, H.X. Hu, V. Filkov, Social synchrony on complex networks. IEEE Trans. Cybernetics **48**(5), 1420–1431 (2018)
21. ESRI—Real estate website, https://www.esri.com/en-us/industries/real-estate/overview
22. Black stone, https://www.blackstone.com/the-firm/asset-management/real-estate
23. J. Wang, Pearson correlation coefficient, in *Encyclopedia of Systems Biology*, ed. by W. Dubitzky, O. Wolkenhauer, K. H. Cho, H. Yokota, (Springer, New York, NY, 2013)
24. The data for our work was taken from: www.terrafly.com/
25. G. Skourletopoulos et al., Big data and cloud computing: a survey of the state-of-the-art and research challenges, in *Advances in Mobile Cloud Computing and Big Data in the 5G Era. Studies in Big Data*, ed. by C. Mavromoustakis, G. Mastorakis, C. Dobre, vol. 22, (Springer, Cham, 2017)
26. Alan Said, Data Science in Practice (Springer Publications, 2019)
27. E. Hart, Biological networks, in *Encyclopedia of Astrobiology*, ed. by R. Amils et al., (Springer, Berlin, Heidelberg, 2014)
28. V. Latora, V. Nicosia, G. Russo, *Complex Networks: Principles, Methods and Applications* (Cambridge University Press, Cambridge, UK, 2017)
29. S. Han, Y. Ko, S. Kim, D.H. Shin, Home sales index prediction model based on cluster and principal component statistical approaches in a big data analytic concept. KSCE J. Civil Eng. **21**(1), 67–75 (2017). Springer publications
30. T. Oladunni, S. Sharma, Spatial dependency and hedonic housing regression model, in: *2016 15th IEEE International Conference on Machine Learning and Applications (ICMLA)*, Anaheim, CA, 2016, pp. 553–558

31. V. Del Giudice, P. De Paola, G.B. Cantisani, Valuation of real estate investments through fuzzy logic. Buildings **7**, 26 (2017)
32. C. Bagnoli, H.C. Smith, The theory of fuzzy logic and its application to real estate valuation. J. Real Estate Res. **16**(2), 169–200 (1998). American Real Estate Society
33. W. Didimo, G. Liotta, F. Montecchiani, Network visualization for financial crime detection. J. Vis. Lang. Comput. **25**(4), 433–451 (2014). https://doi.org/10.1016/j.jvlc.2014.01.002
34. S. Perera, M.G.H. Bell, M.C.J. Bliemer, Network science approach to modelling the topology and robustness of supply chain networks: a review and perspective. Appl. Netw. Sci. **2**, 2–35 (2017). https://doi.org/10.1007/s41109-017-0053-0
35. Z. Wang, J. Han, Visualization of the UK stock market based on complex networks for company's revenue forecast, in *Information and Knowledge Management in Complex Systems. ICISO 2015. IFIP Advances in Information and Communication Technology*, ed. by K. Liu, K. Nakata, W. Li, D. Galarreta, vol. 449, (Springer, Cham, 2015)
36. Realdata, https://www.realdata.com/
37. CREmodel, https://www.cremodel.com/
38. Proapod, http://www.proapod.com/
39. Y. Dong, Value ranges of Spearman's Rho and Kendall's Tau of a class of copulas, in: *2010 International Conference on Computational and Information Sciences*, Chengdu, 2010, pp. 182-185. doi: https://doi.org/10.1109/ICCIS.2010.335
40. S.E. Kumar, V. Talasila, N. Rishe, T.V.S. Kumar, S.S. Iyengar, Location identification for real estate investment using data analytics. Int. J. Data Sci. Analytics, 1–25 (2019)
41. M.J. Moshkov, Time complexity of decision trees, in *Transactions on Rough Sets III*, ed. by J. F. Peters, A. Skowron, (Springer, Berlin, 2005), pp. 244–459
42. D. Hu, Q. Liu, Q. Yan, Decision tree merging branches algorithm based on equal predictability, in: *2009 International Conference on Artificial Intelligence and Computational Intelligence*, Shanghai, 2009, pp. 214–218
43. O.Z. Maimon, R. Lior, *Data Mining with Decision Trees: Theory and Applications*, 2nd edn. (World Scientific, Singapore, 2015)
44. S. Sehgal, H. Singh, M. Agarwal, V. Bhasker and Shantanu, Data analysis using principal component analysis, in: *2014 International Conference on Medical Imaging, m-Health and Emerging Communication Systems (MedCom)*, Greater Noida, 2014, pp. 45–48
45. G.A. Wilkin, X. Huang, K-means clustering algorithms: implementation and comparison, in: *Second International Multi-Symposiums on Computer and Computational Sciences (IMSCCS 2007)*, Iowa City, IA, 2007, pp. 133–136
46. I. Scholtes, Understanding complex systems: when big data meets network science. IT—Information Technology **57**(4), 252–256 (2015). https://doi.org/10.1515/itit-2015-0012
47. A. Bihari, M. K. Pandia, Eigenvector centrality and its application in research professionals' relationship network, in: *2015 International Conference on Futuristic Trends on Computational Analysis and Knowledge Management (ABLAZE)*, Noida, 2015, pp. 510–514
48. F. Grando, D. Noble, L. C. Lamb, An analysis of centrality measures for complex and social networks, in: *2016 IEEE Global Communications Conference (GLOBECOM)*, Washington, DC, 2016, pp. 1–6

Chapter 9
The City of L'Aquila as a Living Lab: The INCIPICT Project and the 5G Trial

Fabio Franchi, Fabio Graziosi, Andrea Marotta, and Claudia Rinaldi

1 Introduction

The success of the fifth-generation (5G) mobile network in Europe is strictly dependent on the creation of a community able to design and develop applications that properly exploit the 5G network potential.

The International Telecommunication Union (ITU) has classified 5G mobile network services into three categories: enhanced mobile broadband (eMBB), ultrareliable and low-latency communications (uRLLC), and massive machine-type communications (mMTC). eMBB aims at meeting the people's demand for an increasing digital lifestyle, and focuses on services that have high requirements for bandwidth, such as high-definition (HD) videos, virtual reality (VR), and augmented reality (AR). uRLLC aims at meeting the expectations for the demanding digital industry and focuses on latency-sensitive services, such as assisted and automated driving, and remote management. mMTC aims at meeting the demands for a further developed digital society and focuses on services that include high requirements for connection density, such as smart city and smart agriculture.

The diversity of use cases for 5G development necessarily translates into a variety of requirements that may be summarized as increased data volumes, number of connected devices, and user data rates together with a reduction of energy consumption, end-to-end latency, and operative costs [1]. Figure 9.1 shows the development lines of the 5G technology.

The existence of a coordination between various pre-commercial solutions of European subjects is the condition for pushing and increasing awareness of the 5G development in Europe. The European Commission has also launched the 5G

F. Franchi (✉) · F. Graziosi · A. Marotta · C. Rinaldi
Università degli Studi dell'Aquila, L'Aquila, Italy
e-mail: fabio.franchi@univaq.it; fabio.graziosi@univaq.it; an-drea.marotta@univaq.it; clau-dia.rinaldi@univaq.it

© Springer Nature Switzerland AG 2020   177
N. V. M. Lopes (ed.), *Smart Governance for Cities: Perspectives and Experiences*,
EAI/Springer Innovations in Communication and Computing,
https://doi.org/10.1007/978-3-030-22070-9_9

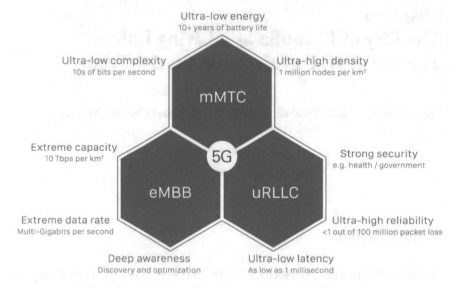

Fig. 9.1 5G technological objectives

Action Plan [2] to stimulate, facilitate, and harmonize initiatives in the European area. The plan envisions a wide large-scale development in 2015, with the perspective of important socioeconomic benefits starting from 2030.

One of the main objectives of 5G trials is to foster an ecosystem around the new 5G capabilities. Vertical industries are involved in the undergoing trial phase that will demonstrate key 5G functionalities and technical/technological enablers. According to 5G PAN European strategy [3], the selection of vertical pilots should take into account sectors including, but not limited to, the media and entertainment, public safety, e-health, automotive, transport, and logistic. Moreover, the strategic role of 5G networks within the main vertical sectors is recognized in the document by 5G-PPP "5G Empowering Vertical Industries" [4]. In order to achieve this aim the experimentation will be carried out in two steps:

1. Until 2018, that is, before the definition of 5G specifications by the Third Generation Partnership Project (3GPP): Tested applications will have to show the 5G technology potentialities, regardless of the state of the standardization process.
2. After 2018, that is, when the first release of 5G specifications will be available: Tested applications will be based as much as possible on standard systems in order to show their interoperability.

It is worth to mention that 5G trials are currently being carried following different streams including private trials, trial events (e.g., UEFA EURO 2020), 5G platforms, and 5G trial cities. Particular relevance in the context of 5G city trials is assumed by the smart cities which have been indicated as one of the pilots of 5G experimentation. Since 5G represents the supporting infrastructure for the Internet of Things (IoT), it will enable new application for smart cities [5]; for example 5G technology will introduce the possibility for the city to govern in real time energy supply, waste management, e-democracy, entertainment events, traffic congestion, and more. Several activities and research projects are currently ongoing regarding 5G applications for the smart city.

The city of L'Aquila is characterized by the presence of a living lab testbed for smart city applications provided by the INCIPICT research project [6] and it is one of the Italian cities hosting the 5G city trials [3]. In this chapter we illustrate the INCIPICT research objectives and provide a description of the smart city testbed which represents a basis for the 5G experimentation. Moreover, we illustrate the use cases that will guide the 5G trial experimentation in the city of L'Aquila.

2   The INCIPICT Project

On the 6th of April 2009, the city of L'Aquila, located in the region of Abruzzo, in central Italy, was hit by a high-magnitude earthquake that killed more than 300 people. The quake destroyed many historical buildings in L'Aquila including medieval churches, basilicas, and cathedrals. Nine years later L'Aquila is an open work-in-progress area that can be used as a test bed for a countless series of experimental projects and the INCIPICT project has been the starting point. This project comes from the conviction that the rebirth of the city of L'Aquila cannot ignore the construction of an innovative communication network that allows citizens and foreign visitors to access information in an easy and efficient way. INCIPICT is intended to build a solid substrate and provide proper interfaces for applications.

The INCIPICT project is focused on the construction of an experimental optical network to build a MAN—metropolitan area network—that consists of an optical ring to connect the main and the most important sites of L'Aquila City. These sites include the strategic node of the University of L'Aquila, in addition to other important sites like public administration (PA) buildings. The optical ring involves the use of some experimental technologies that are designed to test new fiber-optic solutions, energy-saving optical systems, and new networking techniques. This network will also provide the essential substrate for the construction of advanced solutions for the delivery of services such as ultrafast on-demand cloud services, sensor networks, and mobile services.

2.1   The INCIPICT MAN: Metropolitan Area Network

The core of the INCIPICT project is the construction of the experimental optical network for the city of L'Aquila.

Figure 9.2 shows the path of the optical ring through the city, including the axis of the historical center, and highlights the sites of interest along the same.

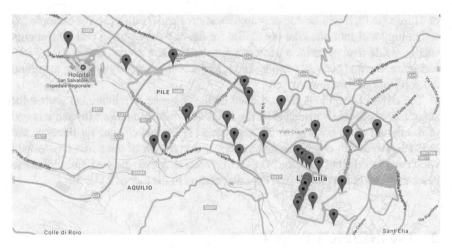

Fig. 9.2 The optical ring and noteworthy sites in red

The optical ring involves the use of some experimental technologies that are designed to test new optical systems in multicore and multimode fibers, space division multiplexing, long-reach Passive optical network (PON) [7], and software-defined optics [8], to name only the main ones. It is conceived and designed as an infrastructure to provide regular operations to the main customers and to serve as a test bed available to the national and the international research community for the investigation and the implementation of innovative communications-related technologies and services. In the long term, the INCIPICT infrastructure guarantees a continuous opening to technological innovations and advanced trials. This makes INCIPICT's experimental network an ideal candidate for the implementation and the testing of the solutions for the 5G mobile network.

Parallely to metro level, the optical and radio access has particular relevance within the INCIPICT project. The University of L'Aquila hosts the INCIPICT 5G Wireless Optical Convergence (WOC) Lab to support research activities in several fundamental aspects of 5G networks such as software-defined optical access, PON front hauling, software-defined control of mobile networks, and multi-connectivity using different radio access technologies (RATs). Facilities and platforms are made available to researchers in order to host a variety of experimental activities and hardware demonstrations for performance evaluations, test, and fine-tuning of 5G advanced techniques. Furthermore, the laboratory will exploit novel aggregation techniques for metro and access network integration. Convergence between different networks will be experimented by utilizing the INCIPICT optical ring.

Future 5G mobile networks will be characterized by centralization of radio access network functionalities in centralized (CUs) and distributed units (DUs). 3GPP has defined eight different functional split options [9], each corresponding to different bandwidths and latency requirements that the so-called front-haul segment should support next-generation optical.

Networks (NGPON2), because of their high capacity, represent a possible supporting infrastructure enabling the front haul of the network. In the context of dynamical instantiation of functional split, also defined as dynamical radio access network (D-RAN), dynamic variation of front-haul segment requirements calls for a wireless-optical convergence in optical network resource management [10].

The dynamism that characterizes the mobile traffic in urban areas, the network requirements on front-haul segment, and the 5G architecture programmability stress on the necessity of developing a control framework based on innovative software-defined mobile network (SDMN) [11]. In 5G converged software-defined access [12], mobile and optical network devices expose, through agents, parameters for the control of the allocation of resources. Our experimentation aims to employ and accordingly extend available open-source platforms [13, 14].

One of the scenarios of interest for 5G is the deployment of dense small-cell networks, namely the ability to install, within an area of interest, a huge number of cells of small size in order to increase network capacity. Such a scenario requires to implement collaborative techniques in order to jointly optimize the radio resource allocation, mitigate interference, and maximize throughput for users. On the other hand, such a tight cooperation impacts the physical infrastructure interconnecting mobile base stations and requires an integrated mobile-optical control [15].

The laboratory will be based on the architecture depicted in Fig. 9.3 with the following features:

Optical access networks realized through the means of both prototyping boards and commercial off-the-shelf OLT and ONU products

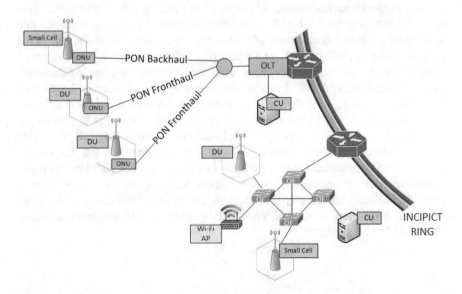

Fig. 9.3 The architecture of the INCIPICT 5G wireless optical convergence lab

An SDN-controlled OpenFlow switch network

An installation of OpenAirInterface (OAI) and EPC with some USRPs for the emulation of small cells, DUs, and CUs

Wireless access points for multi-connectivity experimentation

A software-defined mobile network controller for the experimentation of radio access control techniques

A software-defined optical access control layer for the optimization and control of PONs

An orchestration platform for the optimal deployment of network functions and experimentation of NFV

2.2   Innovative Wireless Technologies

The INCIPICT test bed represents a context for the test of advanced and pervasive wireless technologies, representing a support layer for innovative applications. It will be equipped with a distributed software-defined radio platform composed by 13 nodes supporting 160 MHz instantaneous real-time bandwidth and a 10 MHz–6 GHz frequency range. It provides the opportunity to research on techniques for reduction of energy consumption and increase of transmission speed (and bandwidth). The research program starts from the recent proposal to harness the random behavior of the wireless channel for communication, also called space modulation (SM) [16]. Moreover, research work is focused on network coding (NC) [17, 18]. The NC approach imposes that the nodes must recombine different incoming packets in one or more outgoing packets instead of simply retransmitting what they receive. This approach shows potential benefits compared to normal routing techniques that reside on computational efficiency and robustness of network dynamics. One of the research goals of INCIPICT is to take advantage of the NC to reduce energy consumption of systems and wireless communication networks by adopting the methodology of cross-layer design. Furthermore, an important application of the research in wireless technologies is the real-time tracking and localization [19–21]. People localization allows the creation of healthcare systems with the ability to offer outpatient services. Available technologies for localization such as network or satellite may be utilized in case of emergency to quickly and accurately locate the position of a user and communicate the information as well as for intelligent transportation system applications (connected and autonomous vehicles). Finally, the problem of guaranteeing adequate security levels in wireless networks that are normally characterized by the transportation of sensible data and by limited computational capabilities will be addressed.

2.3   Services and Applications

By having a connectivity at the state of the art for the optical component and for the wireless short-medium-range component as well as an adequate abstraction layer, it will be possible to activate and experiment innovative applications in many areas. In this context the development of experimental services is proposed. The proposed approach is totally open to other application areas in the smart city and community context such as intelligent transportation systems, environment monitoring, homeland security, and energy management.

3   The 5G Experimentation

In order to implement the EC Communication n.2016/588, c.d. "5G Action Plan" by 2020, the Italian Ministry of Economic Development (MISE), with public notice of 16 March 2017, opened the call for project proposals regarding the implementation of pre-commercial experiments for the radio spectrum 3.6–3.8 GHz. Projects should be completed over 4 years in the following geographic areas: (1) Area 1: Milan—metropolitan area, (ii) Area 2: Prato and L'Aquila, and (iii) Area 3: Bari and Matera

5G will be a key enabler in transforming our economy and society by providing proper connectivity in three broad areas: extreme mobile broadband, massive machine communication, and ultrareliable low-latency communication. 5G core will need to run in heterogeneous environments, interacting with multiple types of access network and serving a wide ecosystem of applications and players. End-to-end network slicing will be a key enabler to efficiently manage the diversity of the use cases in the operator network and it should be a relevant factor to be taken into account when designing the 5G core network. Indeed, the overall network architecture should be designed with the capability of guaranteeing the delivering of the proper level of scalability, flexibility, and total cost of ownership (TCO) efficiency.

The use cases mentioned by the European Commission 5G Action Plan confer big relevance to network slicing. Network slicing allows a network operator to provide dedicated virtual networks with functionalities that are specific to the service or customer over a common network infrastructure. The next-generation mobile network should inherently address requirements of such a hybrid network.

Two of the main features of 5G design are cognition and programmability that can be achieved through the so-called softwarization and virtualization techniques for the end-to-end chain of the radio channel, networks, applications, and services.

The new 5G network will also be able to operate in different application contexts (low-power solutions and/or high-reliability, low-latency solutions) providing multilevel network architectures. In this context, classic macro-cell structures coexist and integrate, in a functional manner, with different network types, such as small cell, and they provide different communications mode (relay, device to device (D2D)) to heterogeneous devices (smart objects, cyber physical systems, connected vehicles) related to different requirements in terms of quality of service (QoS).

In the following we illustrate the use cases coordinated by the University of L'Aquila within the context of the 5G trial derived from the INCIPICT project.

3.1 Use Cases

The identification of specific use cases for L'Aquila municipality has been carried out by taking into account the research skills of local universities in the scenario of a smart city and the local vocations of the city. More specifically, the following use cases have been identified:

Structural health monitoring
Building automation and energy management, ICT for cultural heritage
Automotive and the connected vehicle

Figure 9.4 shows the INCIPICT vision of 5G requirements.

The four presented use cases also refer to the INCIPICT project and were originally selected with the aim of introducing innovative technology elements into urban planning and management activities.

3.1.1 Structural Monitoring of Buildings

The project consists of the permanent structural monitoring of buildings and it is focused on the creation of a system of observation, based on minimally invasive technologies, sustainable and innovative, so as to provide information about the structural behavior below normal conditions and in case of seismic events. The large investment of public funds used in the reconstruction of the city of L'Aquila needs an intelligent observatory, able to investigate the effectiveness of the used technologies, especially in relation to their performance over time.

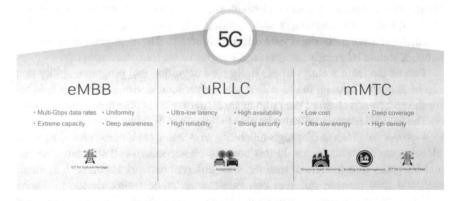

Fig. 9.4 The INCIPICT vision of 5G requirements

The University of L'Aquila has successfully developed and designed a prototype of a system for structural monitoring of a thirteenth-century church, "S. Maria di Collemaggio": a network of wireless sensors equipped with accelerometers, strain gauges, and tilt meters has been implemented and actuated. The system has monitored the dynamic behavior of the church during numerous seismic events occurred after the main shock of April 2009 [22]. Today the structural monitoring system is under deployment on various buildings, trying to catch the structural heterogeneity of the urban scenario. The gathered information will be processed through model-based and data-based approaches, in order to compare and possibly integrate these techniques.

This use case may be considered dependent on the operational scenario: in the case of data collection and processing from sensors in monitored buildings, considering the high number of installed sensors, it can refer to the mMTC context. Vice versa, during a seismic event or just after it, the use case requires high reliability, connectivity, and, sometimes, low latency. Those features refer to the URLL context. It seems interesting to evaluate and experiment the ability of 5G network to dynamically adapt to the changing scenario that this use case can provide.

3.1.2   Building Automation/Energy Efficiency

The focus is on the implementation of a building automation system, in one or more buildings of the University of L'Aquila, to optimize energy consumption and provide early-warning and disaster-recovery systems. This cluster of smart building automation systems will be developed as a living lab, where sensing and actuation data can be read/actuated from external academic and/or industrial entities according to authority levels that are regulated by formal agreements. The monitoring sensor and actuator network, to be developed for the building automation system, will enable the implementation of other systems aimed at protecting the health of citizens in case of natural disasters; for example the ability to constantly track people and the status of the building structure can be exploited to detect the occurrence of dangerous events and activate, when necessary, proper mechanisms for emergency response. Also in this use case, data-based approaches are widely used and they require a huge amount of sensors to be distributed inside and outside the involved buildings. Since this scenario does not require high reliability or low latency it should be referred to the mMTC context.

3.1.3   Enhancement of Cultural Heritage

The virtual reality (VR) application, experienced in loco or remotely, should promote the diffusion of knowledge and places. ICT technologies may be the enabling key to reconnect the virtuality to the reality, thanks to the real-time possibility of interaction. The daily experience should be enriched of more information offered by mobile devices and applications. In particular, augmented reality solutions can provide the opportunity to tell the history of places, showing the transformation of the place over time through computer-based visualizations and multimedia information. For these reasons the development of this kind of applications involves all the layers that characterize the system architecture, from the infrastructural one, through the network one, to the service layer with which the user directly interacts. The use case related to the enhancement of cultural heritage through ICT technologies should have two different behaviors: On the one hand a historical building, a museum, a monument, or an archaeological site may require the deployment of an environmental sensor network for monitoring that defines the use case as mMTC. On the other hand, considering the possibility of extending the tourist experience by exploiting the potential of VR and AR solutions, the same use case is connoted as capable of representing the eMBB segment of the 5G scenario.

3.1.4 Automotive and the Connected Vehicle

The mobility scenario will change significantly in the near future; the new communication technologies, the demand trends, and the need to reduce the environmental impact will require rethinking the mobility paradigm to enable a new driving experience for vehicles connected with transport systems and with the digital world of the future.

In this context, the new geo-localization systems, with the availability of the Galileo satellite constellation (2020), existing (V2X/4G) and emerging (5G) communication technologies, and pressing challenges of cyber security, represent a key element for the future mobility scenario. The challenge of the project is to pursue scientific and technological progress beyond the state of the art in the automotive industry by designing intelligent transport system (ITS) solutions for the transport of people and goods and for the management of emergency situations.

The ITS-equipped vehicle will be aware of the traffic scenario, environmental conditions, and driving context. These data will provide, in advance, all the information needed, for instance, to prevent accidents. For this aim, the ITS will supply the connected vehicles with the latest accurate and geo-localized positioning technologies in order to properly exploit all the benefits coming from advanced dynamic and safety navigation features.

The new capabilities of connected vehicles, such as vehicle cooperation features, introduce security issues which are taken into account in the project. Indeed, another important goal is the correct implementation of the cyber security technology, taking into account that it has not to be an element preventing or degrading the provision of functionalities and services, especially in emergency conditions.

Furthermore, for the sake of functionalities of services (especially safety-related ones), status of channels has to be monitored, to enforce congestion control when needed [23].

4   Conclusion

This chapter presents the INCIPICT project, with particular emphasis on its aim of behaving as a test bed for possible experimentations, especially the 5G experimentation in the city of L'Aquila. The optical network represents the enabling infrastructure for the experimental activities in a living lab context. The test bed will host research activities in optical and wireless transmission as well as optical access/metro transmission and networking for 5G. It has been described how the 5G experimentation in the city of L'Aquila will be driven by suitable use cases for the path towards the future smart city. Four use cases have been illustrated, specifically structural monitoring of buildings, building automation, enhancement of cultural heritage through ICT, and finally connected vehicle. It is worth noting that these are not the only possible exploitations of such an innovative research infrastructure. It has to be underlined that the INCIPICT platform is open to ideas, experimentation purposes, innovative solutions, and technological advances that can be proposed by any research or private entity.

Acknowledgements This research would not be possible without the participation of ZTE Corporation and WindTre (H3G) to the ongoing experimentation process. This work was partially supported by the Italian Government under CIPE resolution no. 135 (December 21, 2012), project INnovating City Planning through Information and Communication Technologies (INCIPICT).

References

1. 5GPPP (2016a), 5G Vision. Whitepaper
2. European Commission (2016), 5G for Europe: an action plan. COM(2016)588
3. 5G Infrastructure Association (2016), 5G PAN-European Trials Roadmap. Whitepaper
4. 5GPPP (2016b), Empowering verticals. Whitepaper
5. K.E. Skouby, P. Lynggaard (2014), Smart home and smart city solutions enabled by 5g, iot, aai and cot services. in: *2014 International Conference on Contemporary Computing and Informatics (IC3I)*, pp 874–878, doi: https://doi.org/10.1109/IC3I.2014.7019822
6. INCIPICT (2017), http://incipict.univaq.it. Accessed: 25 Nov 2017
7. R.P. Davey, D. Grossman, M. Rasztovits-Wiech, D.B. Payne, D. Nesset, A.E. Kelly, A. Rafel, S. Appathurai, S.H. Yang, Long-reach passive optical networks. J. Lightwave Technol. **27**(3), 273–291 (2009)
8. A.S. Thyagaturu, A. Mercian, M.P. McGarry, M. Reisslein, W. Kellerer, Software defined optical networks (SDONs): a comprehensive survey. IEEE Commun. Surv. Tutorials **18**(4), 2738–2786 (2016)
9. 3GPP TR 38801 (2017), Study on new radio access technology: Radio access architecture and interfaces. Release 14
10. A. Marotta, D. Cassioli, K. Kondepu, C. Antonelli, L. Valcarenghi (2018a), Enabling flexible functional split through software defined 5g converged access, in *2018 International Conference on Communication (ICC)*, pp. 1–5
11. T. Chen, M. Matinmikko, X. Chen, X. Zhou, P. Ahokangas, Software defined mobile networks: concept, survey, and research directions. IEEE Commun. Mag. **53**(11), 126–133 (2015)

12. A. Marotta, K. Kondepu, D. Cassioli, C. Antonelli, L.M. Correia, L. Valcarenghi (2018b), Software defined 5g converged access as a viable techno-economic solution, in *2018 Optical Networking and Communication Conference (OFC)*, pp 1–5

13. X. Foukas, N. Nikaein, M.M. Kassem, M.K. Marina, K. Kontovasilis (2016), Flexran: a flexible and programmable platform for software-defined radio access networks, in: *Proceedings of the 12th International on Conference on Emerging Networking EXperiments and Technologies, ACM*, New York, NY, USA, CoNEXT '16, pp. 427–441

14. ONF (2017) CORD—mobile central office re-architected as a datacenter, https://www.open-networking.org/platforms/cord/, accessed 25 Nov 2017

15. A. Marotta, F. Giannone, K. Kondepu, D. Cassioli, C. Antonelli, L. Valcarenghi, P. Castoldi (2017), Reducing comp control message delay in PON backhauled 5g networks, in *European Wireless 2017; 23th European Wireless Conference*, pp. 1–5

16. R.Y. Mesleh, H. Haas, S. Sinanovic, C.W. Ahn, S. Yun, Spatial modulation. IEEE Trans. Vehicular Technol. **57**(4), 2228–2241 (2008)

17. R. Ahlswede, N. Cai, S.Y. Li, R.W. Yeung, Network information flow. IEEE Trans. Inf. Theor. **46**(4), 1204–1216 (2006)

18. S. Katti, H. Rahul, W. Hu, D. Katabi, M. Medard, J. Crowcroft, Xors in the air: practical wireless network coding. IEEE/ACM Trans. Netw. **16**(3), 497–510 (2008)

19. D'Errico L, Franchi F, F. Graziosi, C. Rinaldi, F. Tarquini (2017), Design and implementation of a children safety system based on iot technologies, in: *2017 2nd International Multidisciplinary Conference on Computer and Energy Science (SpliTech)*, pp. 1–6

20. M.A. Hanson, H.C. Powell Jr., A.T. Barth, K. Ringgenberg, B.H. Calhoun, J.H. Aylor, J. Lach, Body area sensor networks: challenges and opportunities. Computer **42**(1), 58–65 (2009)

21. P. Khan, M.A. Hussain, K.S. Kwak, Medical applications of wireless body area networks. Int J Digital Content Technol Appl **3**, 185–193 (2009)

22. V. Gattulli, F. Graziosi, F. Federici, F. Potenza, A. Colarieti, M. Lepidi, Structural health monitoring of the basilica s. maria di collemaggio. Res. Appl. Struct. Eng. Mech. Comput., 823–824 (2013)

23. E. Cinque F. Valentini, A. Iovine, M. Pratesi (2017), An adaptive strategy to mitigate instability in the ETSI DCC: experimental validation, in: *2017 15th International Conference on ITS Telecommunications (ITST)*, pp. 1–6, doi: https://doi.org/10.1109/ITST.2017.7972223

Chapter 10
Mobility in Smart Cities: Will Automated Vehicles Take It Over?

Ralf-Martin Soe

1 Introduction

Many foresight-looking scholars tend to see autonomous vehicles as an inevitable development. For example, Yuval Noah Harari in his book *Homo Deus* compares autonomous vehicles and human-driven vehicles to the horses and human-driven vehicles. Late-nineteenth-century people could not imagine changing their flesh-and-blood emotionally and behaviourally responsive horses to manufactured non-personalised automobiles. According to Harari [1], this switch from horses to cars was an inevitable development as motorised vehicles are significantly more effective and the same outcome will eventually happen to human-driven vehicles that will be replaced by automated vehicles as superior technology. According to innovation researchers (e.g. [2, 3]), this is still not a straightforward process: in some cases superior technologies indeed replace non-superior ones, although it is not automatic and there are several cases, often driven by economic or business reasons, when superior technologies do not make it to the market.

In any way, over the past 100 years, the automobile industry has gone through many incremental and radical innovations but the main concept of a vehicle has remained the same. It is now that the automotive industry is facing one of the biggest revolutions in the history: driverless control of vehicles. There are first elements of computer-assisted driving already in mass production (e.g. adaptive cruise control, parking-assist systems) and there are first city pilots with passenger cars, buses and special-use vehicles.

Therefore, according to growing number of futurists, the main question is only "when" automated vehicles will take over lead in urban mobility. Nevertheless, this chapter is taking more critical approach and allows to ask a question starting with

R.-M. Soe (✉)
Tallinn University of Technology, Tallinn, Estonia
e-mail: Ralf-Martin.Soe@taltech.ee

© Springer Nature Switzerland AG 2020
N. V. M. Lopes (ed.), *Smart Governance for Cities: Perspectives and Experiences*,
EAI/Springer Innovations in Communication and Computing,
https://doi.org/10.1007/978-3-030-22070-9_10

"whether", backed by innovation economics. Therefore, the key question to be analysed in this analysis is whether the transport in future cities will be fully or incrementally autonomous which determines the core setup of future smart cities. In the case of radical change (current human-driven vehicles are like horses that will be opted out from everyday urban traffic by more superior automated vehicles), today's cities' transport systems need to be fully upgraded both physically and virtually—future smart cities would look like the ones in the futuristic movies. On the other hand, in the case of incremental change (e.g. trains, trams and metros being fully automated and open-road vehicles having partially automated functions—something already happening today), autonomous vehicles will take over mainly closed-traffic and offer some automation options for human drivers that still remain in control in open-road traffic—therefore, smart cities of future will still look similar to cities of today.

In other words, this analysis is based on analysing two possible scenarios:

1. **Revolution:** future smart cities will have fully autonomous traffic.
2. **Evolution:** future smart cities will be only incrementally autonomous.

This chapter is set in the following logic. Firstly, autonomous vehicles in the context of smart city concept are analysed. Secondly, barriers and enablers of fully autonomous public urban transport system are evaluated. Thirdly, an empirical overview of cities introducing robot buses follows. Finally, a roadmap for smart city policymakers implementing (or non-implementing) autonomous vehicles is provided.

2   Conceptual Approach

This chapter aims to develop an empirical roadmap for the public sector (mainly cities) on how to implement automated transport in the urban context. This is a theory-driven and empirics-tested approach, meaning that innovation concepts and smart city frameworks will be mapped with examples from the real-life urban cases. It is expected that smart city (and smart mobility) aims to solve actual real-life global problems, and thus United Nations sustainable development goals (SDGs) in the case of urbanisation will be analysed (see Fig. 10.1). As a framework, main drivers and barriers will be analysed in order to provide public sector decision-makers an adequate picture of negative and positive effects of the automated urban transport.

The main research question is to understand under which conditions can we expect the revolution scenario (future cities having fully autonomous transport) and under which conditions can we have an evolution scenario (future cities being incrementally autonomous). Therefore, this is rather a visionary chapter with the aim to analyse what are the conditions (or drivers and barriers) to different scenarios. For this, real-life cases will be analysed, and the empirical part is partially based on the

Fig. 10.1 Conceptual framework

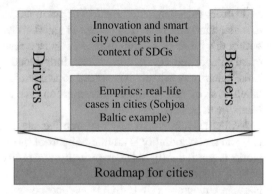

European Union-funded project Sohjoa Baltic (Baltic Sea Interreg project #R073) and its deliverables that tests automated buses in six cities across Europe.

2.1 The Scope and Role of Autonomous Vehicles in the Smart City Concept

This section analyses the role of autonomous vehicles within the innovation economics and smart city domains. In this analysis, autonomous vehicles are seen as potentially superior technology vehicles that aim to replace human-driven vehicles. As mentioned earlier, this process is not automatic. In other words, it is possible that autonomous vehicle is mainly twenty-first-century hype technology with incremental changes (evolution scenario); alternatively, in the case of most extreme revolution scenario, autonomous vehicles are as radical as introduction of personal computers and/or Internet that has significantly changed how we work and live. How do we know which scenario is more probable? For this, we will look into innovation research.

In order to analyse take-up of new technologies, it is important to make a distinction between invention and innovation [3]. Inventions can occur any time but not all of them will be turned into innovations. According to Perez [2], inventions (solutions that are technically feasible) significantly outnumber actual innovations. Innovations need to be economically profitable and socially acceptable before they can widely diffuse. A famous example is BETA technology that was superior to VHS video cassettes but never made it to replace VHS that already dominated the market [4]. Therefore, autonomous vehicles are currently in the status of being inventions that aim to innovate the transport system but their wider success depends on whether they are economically better and socially acceptable. To put it simply, innovations like autonomous vehicles need to have a strong business case and they also need to be widely accepted by people before they can fully diffuse. This will be central to the analysis in the next sections.

In order to understand the effect of automated transport for smart cities, we need
to understand first what the concept of smart cities is, especially the role of mobility
within this concept. Although smart city as a research area is still developing, it is
often categorised via six dimensions, introduced by Giffinger and Haindlmaier [5].
This "smart city wheel" has spread to academia, cities and business sector and is
taken over by the European Commission (see for example the report written by
Manville et al. [6]).

Without doubt, smart city research has clear focus on the smart mobility (e.g.
[7]) and smart mobility is an integral part of most research-based smart city frameworks (e.g. [5, 8–10]). Also empirically, smart city initiatives tend to be often in the
field of mobility and environment (e.g. [6, 11]). In this chapter, we use the United
Nations University (UNU-EGOV)-proposed smart city definition by Estevez et al.
[12], according to which a smart city is:

> "a city which [...] solves multidimensional and complex problems [...] aiming to achieve
> sustainable economic, social and environmental development".

According to UNU-EGOV smart sustainable city framework (see Fig. 10.2),
autonomous vehicles contribute to solving various transport-related problems and
aim to make urban mobility smarter (e.g. via transport systems, accessibility and
infrastructure). Automation of vehicles can also contribute indirectly to the smart
environment domain offering more energy-efficient and economical ways for urban
transit, and is interlinked with other domains.

2.2 Scope of Analysis

It is also important to define the scope of this analysis: In other words—what do we
mean by autonomous vehicles in the urban context? In this analysis we mainly analyse the potential use of automated small buses with a purpose to offer alternatives

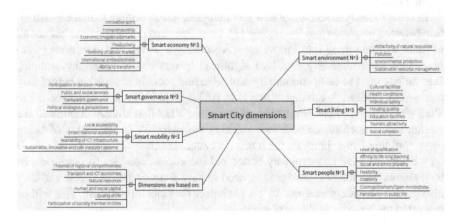

Fig. 10.2 Smart city dimensions by United Nations University

to the last mile transport in cities. This chapter mainly covers the developments of small electric minibuses and their testings on the urban roads, because this is already an ongoing process in many cities globally. In addition, we also track the developments of private cars as they can be used as shared cars in the urban environment.

Cities all over the world are entering the race in terms of who can introduce more robots on their streets. Google launched its self-driving car project in 2009 and has real-life testing experience in California's complex city streets, the average testing mileage reaching over 5000 km per day.[1] Volvo is planning a large-scale test with 100 cars in Gothenburg.[2] Beijing recently announced that it has earmarked 33 road sections with a total length of 105 km for testing autonomous cars.[3] In addition, significant number of cities perform tests with procuring market-ready solutions (such as EasyMile and Navya shuttles) and applying them in urban traffic (with most testing sites in the cities of Europe: Helsinki, Paris, Stockholm, Sion, Toulouse, Wageningen, Lausanne, Tallinn, Trikala, Berlin and others but also in the USA: Texas, Florida and Las Vegas; in Australia: Darwin and South Perth; in Asia: Nanjing, Singapore and Taipei). In addition, there are tens of urban road pilots planned for the next years to come.

In this analysis, it is crucial to distinguish between development stages of automated vehicles, as they apply differently. The most accepted approach among the automated mobility community is proposed by the SAE (Society of Automotive Engineers) International, a global standardisation organisation (see Fig. 10.3). According to SAE, there are the following levels of automation:

SAE 0: NO AUTOMATION (vehicle can provide driving-assist features)
SAE 1: DRIVING AUTOMATION ASSISTANCE (either steering OR braking assisted but not at the same time)
SAE 2: PARTIAL DRIVING AUTOMATION (steering AND braking assisted together as support feature only; human driver must supervise)
SAE 3: CONDITIONAL DRIVING AUTOMATION (automation of full driving task with human fallback; driver must respond promptly when alerted)
SAE 4: CONDITIONAL DRIVING AUTMATION (full automation in predetermined conditions; human must drive when system is not engaged)
SAE 5: FULL DRIVING AUTOMATION (human never has to drive unless he/she wants to)

This chapter does not analyse in depth the use of automated metros, trains, drones, boats, delivery robots, heavy-good vehicles, etc. (grey boxes on graph below) in order to limit the unit of analysis with most commonly spread vehicles globally (cars and buses) which could most radically change the future urban environments when applied on urban roads (see Fig. 10.4). Nevertheless, there are significant developments in these areas. For example, Massachusetts Institute of

---

[1] https://techcrunch.com/2018/04/13/waymo-reportedly-applies-to-put-autonomous-cars-on-california-roads-with-no-safety-drivers/?guccounter=1
[2] https://www.agvegroup.com/volvo-cars-drive-program-100-self-driving-cars-gothenburg
[3] http://www.ecns.cn/business/2018-07-05/detail-ifyvvuhv1809109.shtml?TrucksFoT

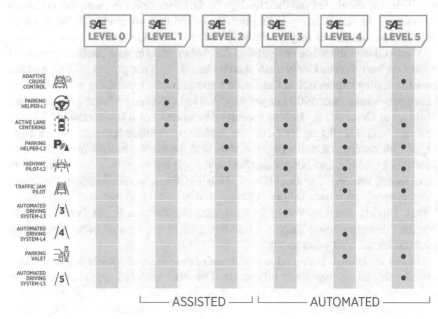

Fig. 10.3 Levels of driving automation. Source: SAE International

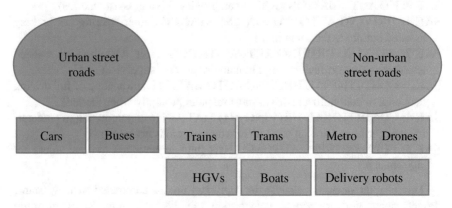

Fig. 10.4 Main focus: cars and buses on urban streets (in blue)

Technology (MIT) and the City of Amsterdam are planning to pilot a fleet of autonomous boats in Amsterdam canals[4] and former Skype founders plan to revolutionise urban deliveries (e.g. food and postal packages) with piloting delivery robots on pedestrian streets in several cities (mainly London, Tallinn and San Francisco).[5]

---

[4] http://senseable.mit.edu/roboat

[5] https://www.starship.xyz

Plus, automation of railroad-based traffic (subways, trams) is already a reality in many cities with the most advantage that city environments can remain largely unchanged.

As this chapter has access to empirical data in planning and conducting actual last-mile automated transport pilots with small electric cars in the six European urban streets, more focus is put to the minibuses (around eight passengers) that could also be seen as hybrid vehicles between buses and cars. These pilots are mainly SAE levels 3–4 with potential to be upgraded to SAE 5. Thus, this analysis focuses on SAE levels 3–5.

2.3   Automated Transport and Sustainable Development Goals

How can we estimate whether autonomous vehicles contribute to the progress or regress of global urban development and whether they contribute to solve or create complex problems, following the UNU smart city definition? One way to tackle this, continuing with UNU-EGOV smart city-proposed smart city framework, is to analyse the globally agreed urban development goals and try to estimate how much automated transport can or cannot influence them. According to the United Nations sustainable development goals (SDGs), one of the global challenges is to make cities inclusive, safe, resilient and sustainable (goal # 11—sustainable cities and communities). Table 10.1 maps UN sustainable goals with the potential of autonomous vehicles.

The effect can be positive (helping to solve the goal) or negative (contributing negatively to achieving the goal). Therefore, it is very crucial to stress that introduction of automated transport is not a linear positive process per se: in some cases, this can have positive consequences whereas weak implementations can also lead to negative consequences. Therefore, it is crucial to analyse the potential positive and negative effects.

3   Barriers and Enablers of Autonomous Public Urban Transport System

In order to answer the question whether autonomous vehicles are radically or just incrementally changing the future, we need to analyse how acceptable they are economically and socially, following the innovation adaption theory by Carlota Perez. This will be pursued via key barrier and driver analysis. If automated vehicles would be introduced fully, urban congestion, traffic accidents and traffic pollution could be minimised, at least theoretically. In other words, there could be less lethal traffic accidents, less traffic jams and cleaner air. According to the Fédération Internationale de l'Automobile (FIA), every day 3500 people die on the roads globally. When

Table 10.1 Mapping of sustainable development goals and automated vehicles

Sustainable cities and communities	The estimated potential of autonomous vehicles
By 2030, ensure access for all to adequate, safe and affordable housing and basic services and upgrade slums	Limited to moderate effect: enhanced mobility improves access to basic services, especially for people living in slums, and reduces the time spent in congestion
By 2030, provide access to safe, affordable, accessible and sustainable transport systems for all, improving road safety, notably by expanding public transport, with special attention to the needs of those in vulnerable situations, women, children, persons with disabilities and older persons	Moderate to very strong effect: fully automated transport could be, at least theoretically, safer, more accessible and sustainable, and there is potential to reduce the costs of transit when implemented full scale In addition, automated transport could also give special attention to the needs of vulnerable situations
By 2030, enhance inclusive and sustainable urbanisation and capacity for participatory, integrated and sustainable human settlement planning and management in all countries	No effect to limited effect: it depends how inclusive and participatory is the process of introducing automated transport for cities. It can help connecting the most vulnerable communities
Strengthen efforts to protect and safeguard the world's cultural and natural heritage	No effect to negative strong effect. Autonomous transport has no effect on protecting the heritage to strong negative effect on non-protecting it (fully automated transport requires reconstruction of urban environments)
By 2030, significantly reduce the number of deaths [...] and decrease the direct economic losses [...] caused by disasters [...] with a focus on protecting the poor and people in vulnerable situations	Limited to moderate effect (both negative and positive). Autonomous vehicles can be designed to be resilient to natural disaster and protect most vulnerable people, but that depends deeply on how they are designed
By 2030, reduce the environmental impact of cities, including by paying special attention to air quality and municipal and other waste management	Up to strong effect. Automated transport can effectively decrease urban congestions and CO_2 emissions by reducing the number of vehicles on cities
By 2030, provide universal access to safe, inclusive and accessible, green and public spaces, in particular for women and children, older persons and persons with disabilities	The effect can range from strong positive to strong negative depending on how the automated transport is implemented. In the positive scenario with smaller number of cars in cities and also parking lots—the access to green and public spaces is enhanced. In the negative cases, reconstructing the cities for autonomous vehicles could also limit this access
Support positive economic, social and environmental links between urban, peri-urban and rural areas by strengthening national and regional development planning	No effect to limited positive or negative effect. This depends on how the automated transport is implemented. If this is planned on the regional levels, then it has positive effect on linkage urban, peri-urban and rural areas. If not, it can introduce mobility silos, thus having a negative effect

(continued)

Table 10.1 (continued)

Sustainable cities and communities	The estimated potential of autonomous vehicles
By 2020, substantially increase the number of cities and human settlements adopting and implementing integrated policies and plans towards inclusion, resource efficiency, mitigation and adaptation to climate change, resilience to disasters, and develop and implement […] holistic disaster risk management at all levels	Limited effect to moderate effect. Depending on how successful the implementation of automated transport is, it can diffuse to limited number of global cities (limited effect) to a majority of cities (strong effect). In any case, automated transport contributed to wiser use of resources
Support least developed countries […] in building sustainable and resilient buildings utilising local materials	No effect

autonomous vehicles take it over, this could be rapidly minimised close to 0, at least theoretically. Automated vehicles coupled with sharing economy concepts would be a very effective measure against large inefficiencies of private cars in cities for two reasons:

1. Significant number of private cars have single driver only.
2. Most of the time, private cars are parked.

According to the much debated OECD [13] study,[6] when autonomous vehicles are coupled with shared economy, nearly the same mobility can be delivered with 10% of the cars exemplified in the case of Lisbon, the capital of Portugal. In other words, it is possible to simulate that one non-stop self-driving shared taxi is as effective in offering mobility as nine private cars in current urban setting. The effect comes from the logic that cars would have more than a single passenger on average and automated cars can be driving non-stop instead of being parked. In the case of larger cities, the MIT Senseable City Lab has modelled earlier based on New York that taxi sharing could reduce the number of trips by 40% with only minimal inconvenience to the passengers [14].

On the barrier side, automated vehicles on public roads require a significant upgrade of urban infrastructure with substantial costs and serious rebuilding involved in already regulated urban environment. Plus, maybe even more importantly, urban citizens would lose a significant fraction of their everyday freedoms such as owning and driving a car for the city traffic purpose and also freedom to take risks and make mistakes in everyday situations (such as exceed speeding limits, cross the street with red lights, park or stop on a wrong place)—driving skill becomes as handwriting skill in a modern computerised office. In this radical scenario, this would have a significant labour market effect as well—number of transport jobs will be effectively substituted by artificial intelligence.

---

[6] https://www.itf-oecd.org/sites/default/files/docs/15cpb_self-drivingcars.pdf

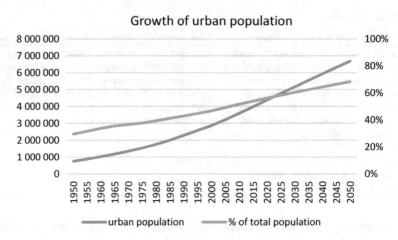

Fig. 10.5 Historical and estimated urban population growth. Source: United Nations. World Urbanization Prospects: The 2018 Revision

3.1 Key Driver: Urbanisation

According to the United Nations World Urbanisation Prospects [15], the urban population of the world has grown rapidly since 1950, having increased from 751 million to 4.2 billion in 2018 (whereas total population, including rural areas, has grown from 2.5 to 7.2 billion). Continuing population growth and urbanisation are projected to add 2.5 billion people to the world's urban population by 2050, with almost 90% of this growth happening in Asia and Africa (see global growth in Fig. 10.5).

Globally, more people live in urban areas than in rural areas, with 55% of the world's population residing in urban areas in 2018. In 1950, 30% of the world's population was urban, and by 2050 68% of the world's population is projected to be urban. More people in cities means also more traffic that needs to be dealt with (e.g. according to Washburn and Sindhu [16], one of the aims of smart city initiatives is to reduce congestion in cities). In the case of urbanisation, significantly more vehicles in cities create real demand for smart mobility solutions. One effective solution is shared automated mobility that can decrease the number of vehicles in cities rapidly.

3.2 Key Driver: Technology

It is very difficult to estimate when and if one technology becomes superior to the current technology and thus has potential to either radically replace previous technology or offer a strong competition. In any case, this is not a momentous situation but it takes significant amount of resources and time to develop new technologies and next it takes time and effort to bring them to the market. For example, landline telephone was replaced by the mobile phones and mobile phones are being replaced by smartphones, although this diffusion takes time and it can be described as a geological sediment where new technologies compete successfully with older ones, although old ones will stay as alternative in place. The same logic applies to TVs and smart TVs, watches and smartwatches, and vehicles. This following subsection analyses the maturity of electric automated minibuses in the case of urban traffic.

The key to understanding autonomous driving is to understand that vehicles need to be equipped with a large number of sensors that provide real-time data to making autonomous decisions. In the case of current minibus pilots, these decision options are pre-programmed, thus making automated minibuses like trams with "virtual trails". When the minibus is put into traffic, their trajectory needs to be recorded several times using sensors and cameras and later the minibus just continues this operation autonomously while reporting continuously on the localisation of the minibus based on analysing sensor and visual data. If everything goes smoothly, minibus can continue this for unlimited number of times without interruption. In the case of unplanned (or unprogrammed) events, the control of the minibus needs to be taken over by human drivers, either via computer on-board or then remotely by computer via Internet. Therefore, the automated minibuses have three ways to control the movement of vehicle:

1. Automated control following the pre-programmed trajectory
2. On-board human control via computer
3. Remote control via Internet and computer

According to Ainsalu et al. [17], automated vehicles are equipped with various sensors that provide data on velocity (based on encoder sensors of mechanical motion that generates digital signals in response to motion) and most importantly geographic position. The vehicles are being constantly localised in real time via a combination of satellites (e.g. global navigation satellite system, GNSS) and odometer sensors. As satellite localisation requires a direct connection between a spot on earth and satellites, in the case of disrupted connection (e.g. being indoors or between tall urban buildings), odometers help to position the vehicle. Automated vehicles also need to constantly report on their exact pose and for this GNSS are integrated often with inertial measurement units (IMUs) that help to measure vehicle orientation using specific sensors (accelerometer, gyroscope and magnetometer). Vehicles have also cameras on-board, and there is an approach to use visual cameras (front and back camera) to improve global positioning (generate a map and estimate robot location based on visual data), although this technique is still developing. The analysis of the surrounding environment is performed via cameras and 3D sensors with main goal to detect objects on the road. The most common sensors are radars in bumper (both front and rear) to detect the distance between vehicle and objects. In addition, automatic minibuses have LIDAR (light detection and ranging) that can estimate the distance between the vehicle and object via emitting a light wave (see Fig. 10.6).

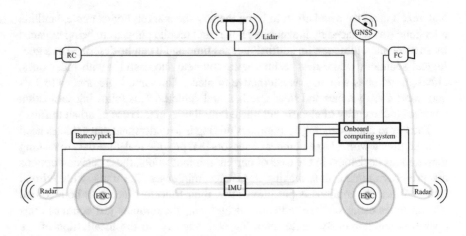

Fig. 10.6 Typical sensors in automated minibuses. Source: Ainsalu et al. (2018) [17]

One can also argue that current-day technology is also a barrier. The sensor technology is far from being as adaptive as expected for real-time decision-making. In this perspective, the current city pilots have experienced the following weaknesses, based on first pilots by EasyMile and Navya minibuses:

1. In the case of unexpected objects (e.g. bicycle passing by), the vehicle is pre-programmed for full stop. Therefore, the speed is often limited to up to 20 km/h in order to avoid any accidents caused by the full stop. The full stop is usually followed by manual on-board control of the vehicle.
2. Strong sensitivity to everyday weather conditions (e.g. drop of rain or falling tree leaves because of wind) too often requires manual takeover.
3. Changes in the visual surrounding environment (e.g. real estate developments) can also stop the vehicle unexpectedly.
4. Sensors and cameras are not developed well enough to follow the simple traffic rules (e.g. there are difficulties in reading the traffic lights and also giving permission to cross the road in the case of pedestrian crossings).
5. In practice, front radar can be too limited, e.g. ignoring higher heavy goods vehicles.

3.3 Key Driver: Market Solutions

Although there are several companies developing market solutions for automated vehicles, there are rather limited options for cities that wish to purchase or lease the buses. Only a limited number of companies (mainly Navya and EasyMile, see Fig. 10.7, but also RDM Autonomous) can deliver buses at this stage, although there are several companies aiming to enter the market soon (e.g. Waymo and General

Fig. 10.7 Main solutions on the market: EasyMile and Navya buses

Motors, Local Motors) or in a few years from now (Apple, Volkswagen Sedric, Robomart). Therefore, the main providers are the following:

EasyMile EZ10. The EZ10 has been the most common demonstrator of autonomous buses. The EZ10 is a battery-powered autonomous electric vehicle designed by Ligier and marketed by EasyMile. It seats up to six people and allows six more passengers to ride standing, or can accommodate a wheelchair. It has been piloted in cities in Finland, Estonia, Norway, the Netherlands, California in the USA and other places. It can operate in metro mode, bus mode and on-demand mode, although the on-demand mode has not been demonstrated on open roads before. With pre-programed routes the on-demand mode can utilise only what has been thought to the vehicle during the programming. EZ10 runs on virtual tracks that are predefined, and needs only light infrastructure to operate (e.g. enough visual cues on road side and possibly local GPS for better positioning). Ligier is a French company.

NAVYA ARMA. NAVYA launched ARMA in October 2015. It is 100% electric and autonomous driverless shuttle that can transport up to 15 passengers and safely drive up to 45 km/h. NAVYA has similar driving modes as EZ10. NAVYA was the vehicle used in the Swiss PostBus demonstration in open roads in 2016 in Sion. NAVYA is a French company.

Local Motors Olli. Local Motors is a US-based company with offices in Europe (Berlin). Olli is a self-driving vehicle designed by Urban Mobility Challenge: Berlin 2030 winner Edgar Sarmiento, and built by Local Motors. The company positions it as "more than a selfdriving vehicle—a platform for new ways of using and thinking around transportation". It demonstrates integration to Watson, and some parts are 3D printed. Autonomy and fleet management are similar to others.

3.4 Key Barrier: Legal Set-Up

From the legal perspective, it is important to distinguish automated driving levels (see Fig. 10.3). In the case of SAE levels 1–2 (driving assistance and partial driving automation), very limited legal innovation is needed, as human still stays fully responsible for driving as today. In the case of SAE 3–5 (conditional driving automation to full driving automation) applied on open streets, significant to radical legal changes are expected. In the longer run, this could even require adding third juridical type throughout the legal system. Currently, driving responsibility (and all other responsibilities) is defined via private or business individuals. In the case of fully automated driving, as is debated, there might be a need of adding robot (or artificial intelligence) individual. In the case of ongoing automated driving pilots applied on open urban streets, very often they are legally constructed as testings of new vehicles, especially when on SAE levels above 3. SAE 5 on urban streets has not been tested; it is legally too complex. In other words, usually SAE3 vehicles on open urban roads have testing licences and human driver as legally responsible. In the case of robot buses, the following laws should be analysed (based on [17]):

- Vehicle registration law (how new vehicles can be registered and put in use; what kind of technical inspections should be carried out and what kind of requirements the vehicle needs to meet; what kind of additional documents should be provided; for example, in Europe, Road Administrations would like to see vehicles following registration regulations in most important parts: how the seats are installed, safety windows, break acceleration, door-closing force, emergency lights, reflectors, lights used in car traffic and where they are installed, kill switch in the bus—based on EU Directive 2007/46)
- Human driver regulations (automated driverless vehicle often cannot obtain a car registration due to its non-compliance regional law (e.g. EU UNECE rules) or local traffic acts; the way to overcome this, in the case of SAE3 vehicles, is to state that every vehicle must have a responsible driver, but in testing automated vehicles the driver can be either inside or outside the vehicle)
- Special testing permit regulations (in many countries, testing automated driving is possible using a test plate certificate, usually up to SAE levels 3–4; these vehicles must have a driver either within the vehicle or acting remotely, who is responsible for the vehicle and takes control of it if necessary; testing can take place on public roads or off-road. Usually testing permissions are given by Road Administrations for a few months, with possibility to extend them; when applying for a test plate certificate, it is often needed to describe how their stewards/ safety drivers will be trained)
- Passenger transportation permit (in the case of involving passengers in testings, there is a need to obtain a taxi or passenger transport permission)
- Driver's licence (due to the need for having a human driver, he/she also needs to have a driving licence; the type of driving licence is determined according to the weight and length of the vehicle as well as the number of passengers; typically, no special licence is needed for automated vehicles)

- Liability law and insurance (one should also follow regulations on product liability, e.g. European Directive 85/374/EEC; the use of automated vehicles within public road traffic up to SAE levels 3/4 raises no special insurance requirements, e.g. traffic liability insurance is a must)
- Criminal law (the main question is whether criminal liability applies only to driver and/or also to manufacturer or any legal entity)

3.5   Key Barrier: Human Acceptance

As mentioned before, innovations need to be also socially acceptable which is often ignored in urban traffic modelling and planning. In 2002, a decision researcher Daniel Kahneman won the Nobel Prize in Economics for a series of work (co-authored with Amos Tversky) proving that people tend to think irrationally, especially in making economical decisions. This chapter is not about rationality analysis but it is clear that the assumption that people are rational is still too often assumed by the traffic, ICT and urban planners, and this can be challenged, similarly to economic decisions. How else can we explain that 3500 people die on roads every day globally,[7] considering that most (if not all) accidents are unintentional—most people do not commute with a purpose to "kill someone" or to "be killed". When using the Kahenman's model (see also Fig. 10.8) people tend to make mistakes when they make intuitive decisions.

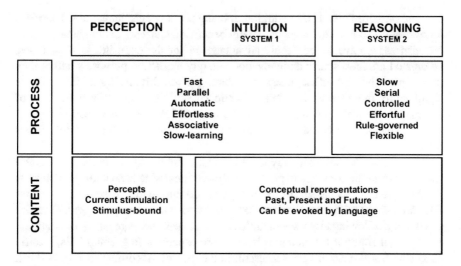

Fig. 10.8 Human rationality and irrationality. Source: Kahneman (2011) [18]

---

[7] https://www.fia.com/3500lives

People-driven urban traffic is full of mistakes. For example, in a midsized European city Tallinn, journalists discovered that 500 traffic mistakes were made during every 60 min in one midsized traffic junction[8] (mainly cars ignoring "no left turn" signs, bus lines, straight lines and also pedestrians crossing the street in wrong place or with red light on); this observation could be continued with stopping and parking mistakes or non-attentive driving (e.g. reading emails on smartphone). This happens in all cities globally every hour with smaller or bigger adjustments.

On the other hand, the logic of autonomous vehicles is that they are programmed to strictly follow all the rules—artificial intelligence or simply a code recorded for driving will most probably avoid at least most mistakes that could lead to unintentional consequences. In other words, it is very probable that autonomous cars and buses would be strictly programmed to actually follow all the traffic rules and this makes it more complicated to have mixed open roads (e.g. both driver-driven and autonomous vehicles on urban roads)—automated vehicles have difficulties in understanding and predicting human behaviour as it often varies and does not follow rules. Therefore, fully rational and automated revolution scenario can take place when opting out human drivers and this might not be socially acceptable. On the other hand, the fact that humans make more mistakes than pre-programmed automated vehicles (at least theoretically) can also make this a strong argument and a driver, once this is accepted socially.

3.6   Key Barrier: Economic Costs

Continuing with Perez [2], in addition to being socially acceptable, novel superior innovations also need to be economically beneficial. Currently, automated vehicles on open roads have no economic advantage, rather the opposite. In most cases, automated minibuses have an innovation and city marketing purpose whereas they do not pass the cost-effectiveness test. Putting a self-driving shuttle in service is significantly more expensive (initiate setup €40–50 00, a monthly rental cost of automated EasyMile or Navya minibus is around €15 000 per bus + costs of drivers' salary) whereas one could lease a human-driven shuttle bus for manifold lower price.

This is also the rationale why most pilots are funded by the competitive R&D funds—it is harder for cities to procure such solutions using local taxpayers' money. At the same time, as the sensor technology is still improving, these buses in operations are capped with speed limit of up to 20 km/h; at the same time regular shuttles can operate following city's speed limits. Nevertheless, there are plenty of empirical examples of superior technologies being more expensive (e.g. smart TVs, electric cars) which still make it very successful in the market. According to Rogers ([19], Fig. 10.9), the key to success is to get a small group of people (innovators) using the

---

[8] http://ekspress.delfi.ee/kuum/autojuhtide-anarhia-tallinna-sudalinna-ristmikul-eiratakse-liiklus-marke-500-korda-tunnis?id=82689315 (in Estonian)

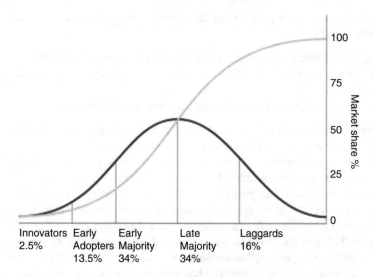

Fig. 10.9 The diffusion of innovations (with successive groups of consumers adopting the new technology (shown in blue), its market share (yellow) will eventually reach the saturation level). Source: Rogers [19]

novel innovations that will be followed by early adaptors, early and late majority and then laggards. Most importantly, innovations do not gain market share linearly but rather exponentially with the first steps being the hardest.

4   Overview of Autonomous Vehicle Initiatives

To date, there are first closed-road pilots conduced all over Europe (e.g. Warsaw, Brussels, Bordeaux, Lean, Trikala, Milan) and globally (e.g. in China[9] and the USA[10]) and first limited trials in the real-life traffic (e.g. in Helsinki and Gelderland). On the other hand, there are no full city-district and no city-level demonstrations and there are no fully automated driving pilots on urban roads (SAE 5 level).

In reality, when zoomed into pilots, evolution scenario tends to be more probable that mimics the development path of technology and regulations. As described in the barrier and enabler section, the current technology is too limited with some fundamental challenges that can actually leave the entry of fully autonomous cars and buses on the urban streets as hype. As the first public tests indicate, the urban

---

[9] https://www.weforum.org/agenda/2018/07/chinese-internet-giant-baidu-has-just-rolled-out-self-driving-buses

[10] http://icities4greengrowth.in/casestudy/pittsburgh-smart-traffic-control-pittsburgh-united-states-america

automated reality is far from letting self-driving cars onto pubic roads with three main weaknesses:

1. Sensor- and image-recognition technology is not advanced enough.
2. All driver's regulation is centred in individual (human) drivers instead of systems/robots.
3. Autonomous vehicles have difficulties in operating in open traffic.

If the first limitation could be seen as technological barrier and second legal barrier which can be at least theoretically solved (or then proven to be non-solvable) then the latter is a fundamental problem that is very difficult to solve without separating autonomous cars and human-driven cars. Namely as first pilots indicate on open roads (e.g. following projects like City2mobile, Sohjoa and others), robot buses tend to be too slow, require manual operators coupled with virtual ones (= driver with a joystick in the bus coupled with a supervisor behind screen) and are inflexible. This has led to the situation that without manual and virtual operators, robot buses in open environment would be stopped in most cases due to either passing-by cars, pedestrians or even small change in physical environment (e.g. close-by construction) or weather conditions (sometimes a drop of rain can stop the vehicle). Therefore, there are no fully automated market-ready solutions for open-road traffic but rather a pre-programed route automation on low speed with actual drivers involved and responsible.

In order to understand the rationale and design of the automated robot bus pilots in the urban contexts, the following section describes one pan-European project and its aims based on the Sohjoa Baltic project documents and deliverables. Firstly, Sohjoa Baltic project team has developed a state-of-the-art report (of which the author of this chapter was one of the contributors, see [17]) listing all known robot bus urban pilots carried out (Fig. 10.10) and also in preparation (see Fig. 10.11). The next section provides in-depth overview of Sohjoa Baltic project for more detailed understanding of how these pilots are planned, mainly based on the project application plan (of which the author of this chapter was one of the contributors).

4.1   Sohjoa Baltic Project

This section is based on the Sohjoa Baltic project proposal and deliverables, co-authored or accessible to the author of this chapter.

The lack of citywide coverage by public transport system increases the automobile dependency for commuters leading to severe congestion on roads, road fatalities, deteriorating air quality and vast CO_2 emissions. Currently public transport is not able to offer competitive option alongside private cars for flexible, on-demand type of operation, and especially the gap in the last-mile connectivity becomes a major barrier to use public transport. The challenge of transition from private cars to public transportation can be addressed by changing the structure of public transport with autonomous operation, introducing safer, attractive, innovative,

Ongoing Pilots in the EU	Duration	Passengers	Travel Distance	Target Group
Bad Birnbach, Germany October 2017–2018	6 months	50 per day	700 m	Inhabitants
Berlin, Germany March 2018	24 months	N/A	1900 m	Inhabitants
Fribourg, Switzerland, 2017	August 2017 onwards	N/A	1300 m	Inhabitants
Château de Vincennes, Paris, France 2017–2018	12 months	200 per day	1880 m	Visitors
Helsinki, Finland 2016–2018	24 months	5600 per pilot	1000 m–3000 m	Inhabitants
La Defense, Paris, France July–December 2017	6 months	N/A	N/A	Inhabitants
Renningen, Germany 2018	N/A	N/A	1200 m	Inhabitants
Saclay, France February–March 2018	2 months	20 per day	2500 m	Inhabitants
Sion, Switzerland, 2016–2018	24 months	60,000 per pilot	1500 m	Visitors
Stockholm, Sweden January–June 2018	6 months	150 per day	2000 m	Inhabitants
Toulouse, France Dec 2017–May 2018	6 months	100 per day	1160 m	Inhabitants
Wageningen, Netherlands 2016–2019	48 months	N/A	200 m–4000 m	N/A
Completed Pilots in the EU				
La Rochelle, Rance, December 2014–April 2015	4 months	14,660 per pilot	1900 m	N/A
Tallinn, Estonia, 2017	3 months	Around 10,000	800 m	Inhabitants and Visitors
Lausanne, Switzerland, 2015	4.5 months	7000 per pilot	1500 m	Inhabitants
Oristano, Italy, July 2014–September 2014	2 months	2580 per pilot	1300 m	N/A
San Sebastian, Spain 2016	3 months	2750 per pilot	1200 m	Inhabitants
Sophia Antipolis, France, February 2016–March 2016	2 months	4059 Per pilot	1000 m	Inhabitants
Trikala, Greece, 2015–2016	3–5 months	12,100 per pilot	2800 m	N/A
Vantaa, Finland, May 2015–August 2015	4 months	19,000 per pilot	900 m	Visitors
Leipzig, Germany, 2016	1–5 months	400 per pilot	1600 m	Inhabitants
Lyon, France October 2016–December 2017	14 months	N/A	1350 m	Inhabitants
Michelin Research Center, France, 2016	6 months	3000 per pilot	1000 m	Inhabitants
Toulouse, France 2017	3 months	3210 per pilot	340 m	N/A
Rest of the World				
Bad Birnbach, Germany October 2017–2018	6 months	50 per day	700 m	Inhabitants
Berlin, Germany March 2018	24 months	N/A	1900 m	Inhabitants
Fribourg, Switzerland, 2017	August 2017 onwards	N/A	1300 m	Inhabitants
Château de Vincennes, Paris, France 2017–2018	12 months	200 per day	1880 m	Visitors
Helsinki, Finland 2016–2018	24 months	5600 per pilot	1000 m–3000 m	Inhabitants
La Defense, Paris, France July–December 2017	6 months	N/A	N/A	Inhabitants
Renningen, Germany 2018	N/A	N/A	1200 m	Inhabitants
Saclay, France February–March 2018	2 months	20 per day	2500 m	Inhabitants
Sion, Switzerland, 2016–2018	24 months	60,000 per pilot	1500 m	Visitors
Stockholm, Sweden January–June 2018	6 months	150 per day	2000 m	Inhabitants
Toulouse, France December 2017–May 2018	6 months	100 per day	1160 m	Inhabitants
Wageningen, Netherlands 2016–2019	48 months	N/A	200 m–4000 m	N/A

Fig. 10.10 List of ongoing and competed robot bus pilots in urban environment. Source: Ainsalu et al. [17]

Adelaide, Australia (N/A)	Oslo & Gjesdal, Norway (2018, 2019)
Calgary, Canada (2018)	San Francisco, U.S. (2020)
Copenhagen, Denmark (2018)	Stavanger, Norway (2018)
Gainesville, U.S. (2018)	Sydney, Australia (2018)
Gothenburg, Sweden (2018)	Christchurch Airport, New Zealand (2018)
Hamburg, Germany (2018)	Melbourne, Australia (2018)
Knoxville, U.S. (2018)	Ann Arbor, Michigan, U.S. (2018)
London, U.K. (N/A)	Shenzhen, China (2018)
Gjøvik, Norway (2018)	Kongsberg, Norway (2018–2019)
Drammen, Norway 2020	Vejle, Denmark (2019)
Tallinn, Estonia (2019, 2020–2022)	Koppl, Austria (2018–2020)
Vienna, Austria (2019)	Helsinki, Finland (2018, 2019, 2020)
Gdansk, Poland (2019)	

Fig. 10.11 List of future pilots in urban environment. Source: Ainsalu et al. [17]

- Pilot in Helsinki
  - 10 months in 2019, starting around March 2019

- Pilot in Tallinn
  - 10 months in 2019, starting around March 2019

- Pilot in Kongsberg
  - 10 months hire of a vehicle. The pilot started September 2018

Fig. 10.12 Large-scale pilots of Sohjoa Baltic project

energy-efficient and improved service. Autonomous transport promotes the usage of urban public transportation including automated driverless electric minibuses as part of the public transport chain especially for first/last-mile trips. Through large-scale pilots in three European cities (see Fig. 10.12) and also 1-month demonstrations in three additional cities (see Fig. 10.15), the project brings institutionalised knowledge and competence on organising environmentally friendly and smart automated public transport solutions as well as providing guidelines on the organisational setup needed for running such a service in an efficient way.

Automated buses will not be optimal everywhere for next few years until technology maturation; therefore Sohjoa Baltic specifically intends to find out first suitable applications and development paths. As all of the development can't be done in laboratories, experiments on the roads are required to bring meaningful data to the discussion. The pilots will act as a proof that the concept is capable to work in transnational environments and can be replicated.

The Sohjoa Baltic project seeks to enhance environmentally friendly transport systems in urban areas by increasing the capacity of urban transport actors, by working out a joint vision, policy and business recommendations as well as short-, medium- and long-term action plan on removing existing barriers and facilitating public transport. These outputs will be used by urban planning authorities, urban transport authorities, companies providing public transport, traffic safety authorities and private sector innovation, service developers and academic and research institutions. This is supported by the increased awareness and improved acceptance of the current and new users of public transportation.

The project aims to provide a toolkit for cities to start the shift towards eco-friendly urban transport. Through the need for developing autonomy and successful paradigm shift from private cars to public transport, traffic will change, emissions will be reduced as well as regional development and consistency will be improved in urban surroundings.

Despite moderately well-executed public transportation, average occupancy of the vehicles in the cities is low, for example about 20% in Finland. In some participant countries (e.g. Poland) the use of public transport has even decreased in the last few years. Instead of mere base traffic, travel chain should be seen as a whole and provide options where the so-called last-mile journey has been resolved. A large part of the traffic between the cities is made with passenger cars because public transportation can't offer competitive alternatives for the last mile. This was proven for instance by the former Kutsuplus service in Finland, which demonstrated that public transport is not able to offer competitive options alongside private cars, even in densely populated regions for flexible, on-demand type of operation.

Automatic vehicles themselves do not solve traffic problems such as traffic congestions and vast CO_2 emissions. Traffic problems can be solved by increasing the modal share of the public transportation (see also Fig. 10.13). The relative efficiency of public transport modes compared to passenger cars is much higher, which means less deteriorated air quality and fewer CO_2 emissions. In addition cars take up a disproportionate amount of space compared to the number of people transported which leads to traffic capacity problems especially in densely populated areas. By using public transportation, more space is released to the housing and parks. Also traffic congestions decrease and traffic safety improves.

As part of the many vehicles featuring self-driving capabilities, automated last-mile public transportation will be among first services. A clear service and cost benefit for automated last-mile public transportation exist, but the products are still developing and slowly entering the market through closed areas such as factories,

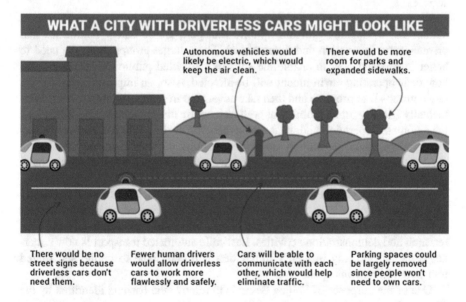

WHAT A CITY WITH DRIVERLESS CARS MIGHT LOOK LIKE

Autonomous vehicles would likely be electric, which would keep the air clean.

There would be more room for parks and expanded sidewalks.

There would be no street signs because driverless cars don't need them.

Fewer human drivers would allow driverless cars to work more flawlessly and safely.

Cars will be able to communicate with each other, which would help eliminate traffic.

Parking spaces could be largely removed since people won't need to own cars.

Fig. 10.13 The potential environmental effect of fully autonomous driving. Source: Chris Dixon; Business Insider

amusement parks and zoos. When technology develops through closed-area operations and open-road pre-commercial pilots, it is evident that the next automation area will be in the last-mile public transportation. For pre-commercial autonomous demonstrations, several countries have the legal framework (such as Finland and Norway, which are also piloting cities in the project) and several are undergoing the legal framework to allow autonomous demonstrations and testing in the traffic (such as Sweden and Estonia). This can later lead to full operation in traffic.

Currently, public transportation is very diverse in many countries, some regions are moving rapidly towards electrified fleets, but some are still running conventional fleets, Norway and Poland as opposed examples. When the use of public transport increases, also the desire and the need to develop the fleet increase. In combination with the growing use of low-emission public transportation vehicles such as biogas and hybrid buses, not to mention the rail traffic, energy consumption savings can be achieved by integrating electric automated last-mile public transportation to the travel chain. The energy needed to operate electric vehicles can be produced completely CO_2 free, depending on the electricity production. Even further energy consumption savings can be achieved by flexible and optimised automated local fleet leading to a truly environmentally friendly urban mobility.

Competitiveness of public transport can be best promoted by increasing the supply, affecting the travel time and reducing prices. Automated operation will change the consistency of public transport, introducing innovative, energy-efficient and improved service. However, to achieve this change, it is necessary to solve the gaps associated to operational, regional, public transportation planning, legal, economical, technological, user acceptance, risk analysis and benchmarking aspects of such services.

Main target groups of the project are urban planning authorities, urban transport authorities, companies providing public transport, traffic safety authorities and private-sector innovation and international. These target groups share the need to better understand how to enable the shift to automated public transportation and how their operating environment will be affected. Also, an important need for the target groups is to promote and then take advantage from the promotion of environmentally friendly urban mobility as well as increase the awareness of how to set up the automated operation and what are the benefits or risks for cost, emissions, service quality, safety, technology and other mobility provider's perspectives. Main need of the users of public transport is to have affordable, yet efficient, public transportation mobility chain service locally.

The European project CityMobil2 (CM2) demonstrated the technical feasibility of automated last-mile transport and fostered the adoption of such new transport systems. From the EU side CM2 has been a milestone on which to build new research and demonstration activities. Last-mile automated transport is now a market issue and the way in which automation will contribute to public and shared transport still remains open.

One of the purposes of Sohjoa Baltic is to remove the barriers identified by the CM2 project.

Barriers like missing marketing and communication strategy to increase the overall acceptance of the automated road transport systems (ARTS), and specifically:

- To increase the level of awareness of the ARTS
- To increase the level of awareness of the benefits of the ARTS
- To correct perceptions that individuals might have for the ARTS in comparison with the conventional transport system

4.1.1 Large-Scale Pilots

Under this group of activities three real-life automated bus pilots will be implemented (Helsinki, Kongsberg and Tallinn). All the pilots will be planned, implemented and evaluated jointly with co-creation activity. Three cities selected for large-scale pilots have unique piloting conditions: all cities have four seasons, including the large variation in daylight (from 6.5 h in December to 19 h in June) and the winters are cold and snowy. The large-scale pilots are meant for the cities where automated bus piloting has already been done (e.g. Finland and Norway) or will be done before the launch of this project.

There has been experiments in Finland, Norway and Estonia. Next the large-scale demonstrations are logical continuation in the selected cities where piloting will be taken to the next level:

1. Automated buses will run for longer period in one location (typically it has been for a day/week but in the case of large-scale demonstrations, it will be at least 1 month in one location). See Tallinn's route in Fig. 10.14.
2. The pilots will be integrated with the city transport network (previous short pilots have been conducted in isolation) and cross-border mobility solutions will be mapped.
3. New mobility options including automated vehicles can be sustained; the next step is to integrate them as part of everyday fleet operated, provided that the outcomes of our pilots are successful.
4. There is effective knowledge sharing between partnering cities, including rotation of operators, if proven necessary.

It should also be noted that large-scale pilots will work in line with small-scale pilots (in Zemgale, Gdansk, Vejle) so that these cities can learn from the best practices.

The large-scale pilots are planned to start late 2018 (Norway) or 2019 (Estonia and Finland) and run throughout the entire year with the following characteristics:

1. Buses will be in operation up to 9 months in each city, mostly in real-life traffic on open roads.
2. Buses will stay in one predefined location for minimum 1 month, so passengers can incorporate them to their everyday mobility plan and create demand for sustaining this service.

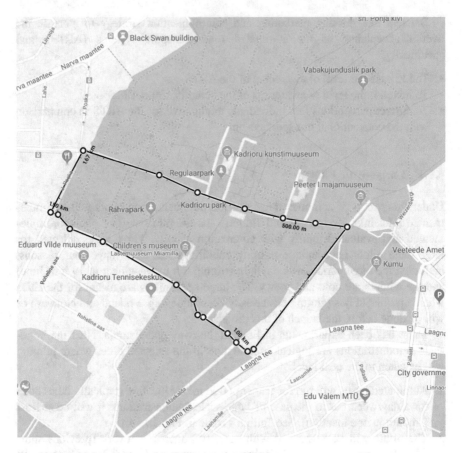

Fig. 10.14 Initial route planned in Tallinn starting 2019

3. During the pilots, buses will be projected to operate up to 10 h a day and 6 days
a week.

The planning, implementation and evaluation of these pilots will be done mutu-
ally in the consortium and this will be effectively documented for other cities'
replications.

4.1.2 Small-Scale Pilots

The small-scale pilots (showcases, see Fig. 10.15) are likely to last for 1 month with
the possibility of setting up more than one bus route within this period. The showcases serve as pilots for transport operators from countries which have no current
legislation on the autonomous transport in place and for the cities where the automated bus piloting hasn't been tested yet. After the large-scale pilots in Helsinki,

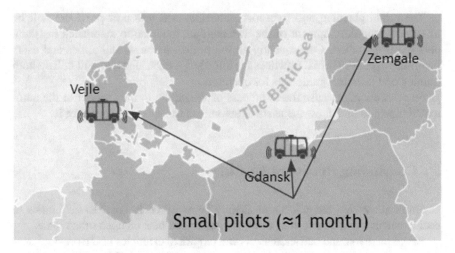

Fig. 10.15 Sohjoa Baltic small pilots

Kongsberg and Tallinn at least three showcases will be organised in Gdansk, Zemgale and Vejle with the following characteristics:

1. While the large-scale pilots test the automated buses in different weather conditions, the showcases raise the awareness of automated transport and have a significant marketing impact.
2. Hosts of the small-scale pilots learn from each other and from the hosts of the large-scale pilots. They work in partnership to plan and organise successful showcases and to evaluate it.
3. To benefit from cross-marketing it is planned to include the small-scale pilots in other significant events, e.g. the European Mobility Week; therefore the buses won't have to be integrated in the city's transport network.
4. The small-scale pilots are planned between August 2019 and February 2020.

Both types of demonstrations will prove that this solution is capable to work in transnational environment and can be replicated:

1. While the small-scale pilots are local, the concept can be implemented in any urban environment that fits the requirements defined in the project.
2. The transnational experience can be extended by a live video streaming of the small-scale pilots from inside of the vehicle so that anyone could be a virtual passenger of the automated bus and the transport providers could better understand the automated intelligent public transport.
3. Hosts of the small-scale pilots will invite local politicians as well as transport providers from other cities and regions in their countries to extend the local character of the demonstrations and share the experience of the automated intelligent transport solutions.
4. Through both large- and small-scale pilots the projects bring competence on provision of the eco-friendly and smart automated transport solutions and guidelines on the logistics and technicalities of running a service.

The urban planning and transport authorities and transport providers will be involved in the development of the transnational roadmap to automated last-mile public transportation. Consequently all the project partners, authorities and users involved will develop best practices for knowledge exchange and will collate both training and technical guidelines for operators.

Small-scale pilots raise the awareness of the public to the concept of the automated transport and allow the users of automated transport to experience it.

5   Concluding Roadmap for Cities

The first automated bus pilots can already be analysed in order to help cities to decide whether and how to start using the automated fleet on open urban roads.

Based on driver and barrier analysis and empirical examples of the previous section a conceptual framework for implementing autonomous vehicles is proposed in Fig. 10.16. The framework encompasses four main components: input, transformation, output and outcome and each of those components is composed by elements. The input component consists of (1) urbanisation, (2) technology and (3) market. The transformation component consists of (4) regulation, (5) cultural and (6) economic. The component output consists of (7) evolution and (8) revolution and finally the component outcome consists of achievement of the (9) SDGs.

The framework is directed at policymakers on three levels:

- Local government (e.g. cities)
- Central government (central and federal governments)
- Regional government (multiple areas/countries)

It should be noted that the aim of this chapter is not to predict the future; it is rather to analyse under which conditions future cities can have fully automated vehicles on public urban roads, that is, how the revolution scenario can realise. On the other hand, there are clear reasons to argue that the evolution scenario is more probable as fully automated transport assumes radical rebuilding of cities. Nevertheless, both scenarios are possible, although their probabilities are rather dynamic, changing across time and locations. In other words, the probability to introduce automated fleet is expected to increase over time and applies more to novel or more adaptive cites.

There are clear incentives for introducing automated vehicles on open urban roads. Firstly, in the area of urbanisation, the cities just need to cope with increased

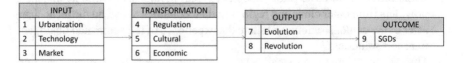

Fig. 10.16 Framework to implement autonomous vehicle initiatives

congestion and thus automated and shared urban transport can effectively traffic smoother, reducing the need of vehicles up to ten times for the same number of trips. Secondly, there is already more than a decade of technology advancement which has intensified over the past years. The breaking point will be when automated vehicles prove to be superior to human driven. Thirdly, already first market solutions like EasyMile and Navya vehicles are available for all cities, and the demand tends to be higher than supply.

As explained in the innovation's entry barriers to the market, there are also examples when superior technologies do not reach mainstream, mainly due to social non-acceptance and economic costs. Therefore, there are also lock-ins or disincentives that can block automated vehicles from becoming a mainstream: firstly, the current legal system that needs to be upgraded with third legal person: artificial intelligence; secondly, humans as collective life freedoms related to open traffic and human driving and there can be logical resistance towards giving this ability to drive away to robots; and thirdly, automated vehicles need to get much cheaper compared to the current solutions on the market.

To conclude, the aim of cities pursuing to become smart should not be technology driven but should be to help solving actual global challenges on the broader level. Thus, this analysis maps United Nations smart sustainable goals with potential and threats of automated vehicles in the case of urban development.

Acknowledgements This work has been supported by the European Commission through the Interreg Baltic Sea Region project Sohjoa Baltic (#R073) and it is also linked to the project "Strengthening Governance Capacity for Smart Sustainable Cities" co-funded by the Erasmus+

References

1. Y.N. Harari, Homo deus: a brief history of tomorrow, 2016
2. C. Perez, Technological revolutions, paradigm shifts and socio-institutional change, in *Globalization, Economic Development and Inequality: An alternative Perspective*, 2004, pp. 217–242
3. J. Schumpeter, *Business Cycles* (McGraw-Hill, New York, 1939)
4. W.B. Arthur, Competing technologies: an overview, in *Technical Change and Economic Theory*, ed. by G. Dosi et al., (Printer Publishers, London and New York, 1988), pp. 560–607
5. R. Giffinger, G. Haindlmaier, Smart cities ranking: an effective instrument for the positioning of the cities? ACE: Architecture City Environ. **4**(12), 7–26 (2010). Retrieved from http://upcommons.upc.edu/revistes/handle/2099/8550
6. C. Manville, G. Cochrane, J. Cave, J. Millard, J.K. Pederson, R.K. Thaarup, A. Liebe, M. Wissner, R. Massink, B. Kotterink, *Mapping smart cities in the EU* (2014), p. 200. https://doi.org/10.2861/3408
7. M. Batty, K.W. Axhausen, F. Giannotti, a. Pozdnoukhov, a. Bazzani, M. Wachowicz, Y. Portugali, Smart cities of the future. Eur. Phys. J. Spec. Top. **214**(1), 481–518 (2012). https://doi.org/10.1140/epjst/e2012-01703-3
8. H. Chourabi, T. Nam, S. Walker, J.R. Gil-Garcia, S. Mellouli, K. Nahon, H.J. Scholl, Understanding smart cities: an integrative framework. in *Proceedings of the Annual Hawaii International Conference on System Sciences*, 2011, pp. 2289–2297. doi:https://doi.org/10.1109/HICSS.2012.615

9. T. Nam, T.A. Pardo, Conceptualizing smart city with dimensions of technology, people, and institutions, in *Proceedings of the 12th Annual International Digital Government Research Conference on Digital Government Innovation in Challenging Times - Dg.o '11*, 282, 2011. doi:https://doi.org/10.1145/2037556.2037602

10. P. Neirotti, A. De Marco, A.C. Cagliano, G. Mangano, F. Scorrano, Current trends in smart city initiatives: some stylised facts. Cities **38**, 25–36 (2014). https://doi.org/10.1016/j.cities.2013.12.010

11. A. Caragliu, C. Del Bo, P. Nijkamp, Smart cities in Europe. J. Urban Technol. **18**(2), 65–82 (2011). https://doi.org/10.1080/10630732.2011.601117

12. E. Estevez, N. Lopes, T. Janowski, Smart sustainable cities: reconnaissance study, 2016, pp. 1–330.

13. OECD, Urban mobility system upgrade. How shared self-driving cars could change city traffic? 2015., Retrieved from: http://www.itf-oecd.org/urban-mobility-system-upgrade-1

14. P. Santi, G. Resta, M. Szell, S. Sobolevsky, S. Strogatz, C. Ratti, Taxi pooling in New York City: a network-based approach to social sharing problems, 2013. https://arxiv.org/abs/1310.2963

15. UN: World Urbanization Prospects (2014). Retrieved from: https://esa.un.org/unpd/wup/publications/files/wup2014-highlights.Pdf

16. D. Washburn, U. Sindhu, Helping CIOs Understand "Smart City" Initiatives. Growth **17**, 1–7 (2009)

17. J. Ainsalu, V. Arffman, M. Bellone, M. Ellner, T. Haapamäki, N. Haavisto, E. Josefson, A. Ismailogullari, E. Pilli-Sihvola, O. Madland, R. Madzulaitis, J. Muur, S. Mäkinen, V. Nousiainen, E. Rutanen, S. Sahala, B. Schønfeldt, P. Smolnicki, R. Soe, J. Sääski, M. Szymańska, I. Vaskinn, M. Åman, State of the art of automated buses. Sustainability **10**, 3118 (2018)

18. D. Kahneman, *Thinking fast and slow* (Farrar, Straus and Giroux, New York, 2011)

19. E.M. Rogers, *Diffusion of innovations*, 5th edn. (Free Press, New York, NY, 2003)

Index

A

Air pollution
  Climo, 141
  decision-making, 142
  IoT, 142
  sphere of energetics, 141
  support system, 142
  transport, 141
  WSN, 142
Amsterdam Innovation Motor (AIM), 34
Annual Budget Plan, 84
Architectural disability, 78
Artificial intelligence (AI), 115
Automated road transport systems
        (ARTS), 211
Automated transport, 196
Autonomous public urban transport system
  automated vehicles, 197
  current urban setting, 197
  infrastructure, 197
  key drivers
    economic costs, 204, 205
    human acceptance, 203, 204
    legal set-up, 202, 203
    market solutions, 200, 201
    technology, 198–200
    urbanisation, 198
Autonomous vehicle initiatives
  physical environment, 206
  public tests, 205
  technological and second legal
        barrier, 206
  technology and regulations, 205
  urban environment, 207

B

Barcelona City Council, 35
Barcelona Open Government, 35
Base Erosion and Profit Shifting (BEPS), 119
Basic regression methods, 153
Blockchain technology
  cryptocurrency, 104, 106
  cryptography, 104
  financial asset, 104
  financial resources, 105
  financial transactions, 105
  information exchanges, 107
  intrinsic characteristics, 107
  lack of illegality, 105
  legal nexus, 106–108
  personal criterion, 108
  quantitative criterion, 108
  regulatory framework, 106
  social and economic policies, 105
  spatial criterion, 107
  technological innovation, 106
  temporal criterion, 107
  virtual and international business field, 105
BreezoMeter Ltd. Company, 143

C

Centralized units (CUs), 180
Centre for Smart Cities, 76
City as a Personality, 56
City marketing, 56, 58, 62, 66, 69
CityMobil2 (CM2), 210
Civil society, 86
Closed-circuit television (CCTV), 143–144

© Springer Nature Switzerland AG 2020   217
N. V. M. Lopes (ed.), *Smart Governance for Cities: Perspectives and Experiences*,
EAI/Springer Innovations in Communication and Computing,
https://doi.org/10.1007/978-3-030-22070-9

Conflict situations, 148
Cyberpunk, 87
Cyberspace Administration of China (CAC),
       97, 103

D
DataFromSky System
    application, 148, 149
    areas of use, 146, 147
    principle of operation, 144, 145
Department of Economic and Social Affairs,
       104, 112
Deputy Commissioners, 23
Development Policy and Analysis, 104
Digital economy
    access to technology, 84
    analytical positioning
        basic financial services, 96
        Bitcoin, 92, 93
        blockchain technology and
            cryptocurrencies, 102
        double spending, 94
        economic policy events, 100
        economic strategy, 100
        economic transaction, 98
        education and partnerships, 99
        encryption, 95
        financial business, 100
        financial institution, 94, 102
        financial instruments, 102
        financial resources, 93
        financial services, 96
        financial systems, 93, 100
        global economic flow, 92
        government and financial supervisory
            authority, 103
        humanity, 92, 99
        hypothetical-conditional form, 101
        innovative socioeconomic system, 100
        interconnection and communication, 97
        international legal community, 101
        IP-based networks, 97
        IPv6-based networks, 97
        legal and socioeconomic doctrine, 98
        legal behavior, 102
        legal doctrine, 92
        legal immigrants, 99
        legal payment instrument, 103
        logical method, 101
        logical-semantic constructivism, 102
        mining, 95
        payment methods, 103
        popular pressure, 100

        regulatory framework, 94, 102
        social event, 93, 101
        technological innovation, 92
        world economic integration, 97
    artificial intelligence (AI), 115
    blockchain and cryptocurrencies, 84
    civil society's depoliticization process, 113
    commercial relationship, 111
    development of education, 112
    disruptive technologies, 84, 85
    economic and sustainable development, 110
    economic shocks, 109
    economic transaction, 116
    flow of economic growth, 116, 118, 119
    human evolution, 110
    human rights, 109
    infrastructure and installations, 125
    international organizations, 113
    Internet governance and regulatory
            framework, 85
    learning machine and external resources, 115
    legal doctrine, 110
    logical-semantic constructivism, 85
    philosophical and sociological studies, 85
    political globalization, 111
    public policy, 109
    regulatory framework, 85, 128
    rule of law, 110
    smart economy, 116
    social and economic benefits, 111
    social and financial aspects, 84
    social awareness, 111
    social-economic event, 84
    social fact and constructive process
        attributes, social fact, 87
        autonomy and social rights, 86
        competent legal language, 88
        currency, 87
        cyberpunk, 87
        democratic system and collective
            consciousness, 88
        disruptive technological age, 91
        economic and financial system, 90
        emergence of cryptocurrency, 88
        international economic scenario, 91
        law science, 89
        layer of protection, 91
        logical schemas, 88
        logical-semantic constructivism, 88
        protection and progress, 90
        social and economic reality, 91
        social reaction, 89
        socioeconomic event, 88
        virtual communication technology, 89

social justice, 113
social observation approaches, 85
social work, 114
sovereignty and border protection
    bank system, 120
    cloud environment/virtual economy, 121
    financial system, 121
    leverage process, 120
    policy, 120
    sector, 120
    smart economy, 121
state administration and social programs, 128
sustainable development, 112
tax law and international policies, 116
tax morality, 116, 118, 119
Distributed units (DUs), 180
Domain name system (DNS), 90

E
Electronic Commerce Act, 122
Enhanced mobile broadband (eMBB), 177
Environmental Protection Agency (EPA), 137
European Commission, 192
European Environment Agency (EEA), 138
European Industrial Revolution, 110

F
Federal Revenue Service, 104, 105
Fédération Internationale de l'Automobile
        (FIA), 195
Fifth-generation (5G) mobile network
    city of L'Aquila, 179
    classic macro-cell structures, 183
    diversity of, 177
    economy and society, 183
    ecosystem, 178
    end-to-end network slicing, 183
    European Commission, 177
    geographic areas, 183
    INCIPICT research project, 179
    latency-sensitive services, 177
    network slicing, 183
    next-generation mobile network, 183
    requirements, 177
    smart city and smart agriculture, 177
    softwarization and virtualization
            techniques, 183
    technological objectives, 178
    use cases
        automotive and connected vehicle, 186
        building automation/energy
                efficiency, 185

enhancement, cultural heritage,
        185, 186
    INCIPICT vision, 184
    L'Aquila municipality, 184
    structural monitoring, buildings, 184
    vertical industries, 178
Finance Department and Finance Minister, 22
Florida International University (FIU), 155, 157
French National Assembly, 86
French Revolution (1789–1799), 86

G
General Data Protection Regulation
        (GDPR), 114
Global navigation satellite system
        (GNSS), 199
Gross national product (GDP), 108

H
Highest magnitude win approach, 161
House price index (HSI), 156
Human health, 135

I
INCIPICT Project
    experimental optical network, 179
    high-magnitude earthquake, 179
    innovative communication network, 179
    MAN, 179–181
    open work-in-progress area, 179
    public administration (PA) buildings, 179
    services and applications, 183
India's Gujarat International Finance
        Tech-City (GIFT smart city), 19
Inertial measurement units (IMUs), 199
Information and communication technologies
        (ICTs), 7, 31, 33, 112, 115
Initial coin offering (ICOs), 120
Institutional banking transactions (IOF), 107
Intelligent transport system (ITS) solutions, 186
International inclusive framework, 85
International tax law, 84
International Telecommunications Union
        (ITU), 3, 177
Internet of Things (IoT), 19, 178–179
Istanbul IT and Smart City Technologies Inc.
        (ISBAK), 8
Istanbul Metropolitan Municipality (IMM), 8
Istanbul Smart City Index Study, 9
Italian Ministry of Economic Development
        (MISE), 183

K
Key performance indicators (KPIs), 7

L
Lahore Metropolitan Authority, 23
Light detection and ranging (LIDAR), 199
Linear regression, 156
Local Government Fund, 22

M
Machine learning techniques, 153, 154, 156,
     157, 159, 164, 174
Mars Group Workshop, 9
Massachusetts Institute of Technology (MIT),
     112, 193–194
Massive machine-type communications
     (mMTC), 177
Matrix rule of tax incidence (MRTI), 89, 101
Metropolitan area network (MAN), 179–181
Metropolitan Smart City Authority, 25, 26
Mobypark applications, 136
Motor Vehicle Emission Simulator (MOVES
     system), 137
Multiple listing service (MLS), 158

N
National Survey of Student Engagement, 85
Network coding (NC), 182
Network science
     attribute and landmark, 165–169
     attributes, 153
     attribute selection, 165
     bipartite network, 155, 165
     condominiums of Alton Rd, 171
     condominiums of James Ave, 170
     data analytics, 154
          attribute selection, 159, 160
          decision trees, 161
          K-means clustering, 162, 163
          PCA, 162, 163
     data analytic tools, 154
     data set, 157, 158
     deviation error, centroids, 170
     dwellings and commercial property, 154
     eigen centrality values, 171
     hedonic modeling, 153
     identification, 153
     K-means clustering, 165
     link weight change, 172
     location identification, 155, 163, 164

     machine learning techniques, 155
     methods and tools, 153
     mutual influence, 154
     Pearson correlation coefficient, 154
     principal component analysis, 165
     public–private partnership, 153
     related works, 156, 157
     requirement, 153
     scopes of, 155
     smart governance, 155, 156, 172, 173
     social media, 154
     state-of-the-art comparison, 156, 157
     user's requirements, 154
     values, attributes, 170
New Bus Network project, 35
Non-governmental organisations (NGOs), 9

O
OECD Model Tax Convention, 119
Ontology Based Information Extraction
     (OBIE), 32
Ontology Design Patterns (ODP)
     goal-objective-strategy, 44
     mission-goal, 44
     pre-design phase, 39
     problem-cause-effect, 42
     problem-solution-vision, 42, 43
     project, 46
     strategy, 44, 47
     strategy ontology, 39
     Web-based platform, 39
Organisation for Economic Co-operation and
     Development (OECD), 98, 119

P
Pakistan, smart cities
     characteristics, 18
     cities' governance, 22–24
     educational institutions, 17
     policy documents, 17
     policy planners, 17
     proposed model, 23, 25, 26
     wireless network sensors and
          e-connectivity, 17
Passive optical network (PON), 180
Portable emission measurement system
     (PEMS), 139
Post-encroachment time (PET), 148
Principal component analysis (PCA), 162, 163
Provincial Steering Committee, 26
Public urban transport, 190

Q

Quality of life (QoL), 33

R

Radio access technologies (RATs), 180
Real driving emission (RDE), 138
Real estate investment, 154, 157, 159, 163, 172, 174
Robert Bosch GmbH company, 141

S

Seven Principles of Universal Design, 78
Sixth dimension of human rights, 84, 85, 112
Small- and medium-sized enterprises (SMEs), 17
Smart Air Quality System, 143
Smart cities
    competitiveness and sustainability, 76
    critical discussion, 80
    design buildings and infrastructure, 4
    design for all, 78, 79
    disabilities, 73
    goals and sustainability, 3
    international agreements, 3
    interpretations, 76
    methodologies
        Big Smart Istanbul, 8, 9
        comparative analysis, 11, 14
        Dubai, 6, 7
        Montreal Smart and Digital City, 10, 11
    people, special needs, 76, 77
    publications, 4
    research methodology, 5
    scientific databases, 4
    social and economic inequalities, 73
    urban agendas and policies, 76
    urbanisation and growth, 73
    UNU-EGOV, 3
    world population, 73
Smart Citizen Kit, 34
Smart City Steering Committee, 24–26
Smart city strategy development
    Amsterdam, 34
    Barcelona, 34
    comparison and step identification, 37
    corresponding ontologies, 48
    data collection and knowledge extraction, 32, 33
    Edmonton, 35
    effective and efficient development, 30
    generalized development process, 37, 38

    identification of, 30
    London, 36
    New Castle vision, 35
    ontology, 30
    planning context, 36
    platform design and implementation, 47
    related work and problem investigation, 31, 32
    research method process definition, 30
    set of ontology design patterns, 30
    use case, 50, 52
    validation of, 39, 40
    Vienna, 33
    vocabulary, 33, 34
    Web-based platform, 30
Smart City transformation, 17
Smart City Wien Framework strategy, 34
Smart contracts
    agency-centric model, 122
    economic flow, 123
    economy and power, 124
    financial system, 123
    financial transactions and payments, 122
    human reality, 123
    human security, 124
    international cooperation, 124
    legal entity, 123
    legal nature encompasses technological resources, 121
    legal personality, 122
    physical borders, 124
    regulatory system, 122
    self-executing, 121
    sense of global community, 125
    stability of, 124
Smart governance
    attributes and requirements, 22
    community, 18
    conceptualization, 19, 21
    decision-making processes, 21
    definition, 18
    diversity, 18
    domains of confusion, 20
    literature, 19
    politics, 20
    service delivery, 21
    smart administration, 19
    smart city transformation, 21
    sociopolitical context, 20
    Spanish smart city, 19
    type of, 19
    typology, 18
    UNU-EGOV, 22

Smart governance (*cont.*)
  urban collaboration, 20
  urban development issues, 19
Smart mobility
  automobile industry, 189
  autonomous vehicle initiatives, 214
  computer-assisted, 189
  conceptual approach
    automated transport and sustainable
      development goals, 195, 196
    public sector, 190
    scope and role, autonomous vehicles,
      191, 192
    scope of analysis, 192, 193
    theory-driven and empirics-tested
      approach, 190
  human-driven vehicles, 189
  incremental change, 190
  innovation economics, 190
  innovation researchers, 189
  lock-ins/disincentives, 215
  policymakers, 214
  radical change, 190
  revolution and evolution, 190
Social media
  Bratislava, 61, 62
  Bratislava—hlavné mesto SR, 64, 66
  Citizen's Interest Groups/Urban Activists,
    61, 62
  city marketing, 59–61
  commercial business practice, 56, 58, 59
  communication channels, 56
  comparative analyses, 68
  complexity and hierarchy, 56
  corporate identity, 56
  elements, city identity, 56
  ethical communication, 57
  European cities and regions, 55
  Facebook page, 69
  infrastructure fuel sustainable
    development, 56
  Ivo Nesrovnal pre Bratislavu, 66
  learning system, 58
  living organism and communication, 57
  methodology, 62–64
  municipal learning, 70
  principal task, 70
  process of communication, 70
  psychological needs, 58
  SMART city, 56
  smart/intelligent, 58
  SMART project/initiative, 55
  social cohesion, 58

  soft factors, 55
  specific characteristics, 57
  strategies and planning, 69
  system/city, 58
  transmitting, 57
  unique organization, 56
Software-defined mobile network (SDMN), 181
Sohjoa Baltic project
  automatic vehicles, 209
  autonomous transport, 208
  closed-area operations, 210
  groups of project, 210
  large-scale pilots, 208, 211
  open-road pre-commercial pilots, 210
  potential environmental effect, 209
  public transportation, 209, 210
  public transport system, 206
  small-scale pilots, 212–214
  transnational environments, 208
  transport systems, 208
Space modulation (SM), 182
Special-needs approach, 78
Stationary Sources Modelling System
    Methodology (SYMOS'97), 137
Strategy and Action Plan, 10
Strategy development
  decision makers, 29
  stakeholders, 29
  social, environmental and economical
    challenges, 29
Strengths, Weaknesses, Opportunities and
    Threats (SWOT), 9
Supply chain networks, 157
Sustainable development
  accessibility, 75
  homogenous group, 74
  mobility, 75
  public space, 74
  social interaction, 74
  urban safety, 74
Sustainable development goals (SDGs), 3,
    190, 195, 196
Sustainable transport
  agricultural production, 135
  air pollution and environmental
    pollution, 135
  American MOVES model, 138
  automobile transport, 135
  cameras, 139
  Copert, 138
  efficiency and service, 136
  emission values, 137, 139
  e-trailer, 139
  EU Commission, 136

German company PTV Group, 137
human health and environment, 135
and infrastructure, 136
low-emission zones, 140
measurement, 139
mobile sources, 136
monitoring emissions, 136
motor vehicle emissions, 137
PTV Vissim, 137
PTV Vistro, 137
PTV Visum, 137
PTV Viswalk, 137
public transport, 136
street-level emission map, 138
traffic density, 139
traffic images, 139
vehicle types, 139
Sustainable world, 84, 113–116

T
Tax on the transmission of assets by death or
    donation (ITCMD), 105
Technology Arrangements Service Bill (TAS
    Bill), 121
TerraFly, 155
Test bed, 179, 180, 182, 187
Third Generation Partnership Project
    (3GPP), 178
Time exposed time to collision (TET), 148
Time to collision (TTC), 148
Transport Research Centre, 139
Transport system, 191

2030 Agenda for Sustainable Development
    from Department of Economic and
    Social Affairs in the United Nations,
    92, 97, 112, 126

U
Ultrareliable and low-latency communications
    (uRLLC), 177
United Nations Conference on Trade and
    Development (UNCTAD), 118
United Nations University (UNU-EGOV), 192
Universal Declaration of Human Rights, 98
Urban Ecology Agency, 35

V
Vehicle2Grid, 34
Vienna Smart City Strategic Plan, 32
Virtual counting gate, 148
Virtual reality (VR) application, 185

W
Web 2.0, 58–61
Web of Science and Google Scholar, 18
Wireless optical convergence (WOC), 180
Wireless sensor network (WSN), 139, 142
World Bank Financial and Private Sector
    Development Consultative Group, 96
World Bank Group (WBG), 91
World Food Program (WFP), 104
World War I (1914–1918), 86

Printed in the United States
By Bookmasters